The Open University

LA4

Linear transformations

This publication forms part of an Open University course. Details of this and other Open University courses can be obtained from the Student Registration and Enquiry Service, The Open University, PO Box 197, Milton Keynes, MK7 6BJ, United Kingdom: tel. +44 (0)870 333 4340, e-mail general-enquiries@open.ac.uk

Alternatively, you may visit the Open University website at http://www.open.ac.uk where you can learn more about the wide range of courses and packs offered at all levels by The Open University.

To purchase a selection of Open University course materials, visit the webshop at www.ouw.co.uk, or contact Open University Worldwide, Michael Young Building, Walton Hall, Milton Keynes, MK7 6AA, United Kingdom, for a brochure: tel. +44 (0)1908 858785, fax +44 (0)1908 858787, e-mail ouwenq@open.ac.uk

The Open University, Walton Hall, Milton Keynes, MK7 6AA.

First published 2006.

Edited, designed and typeset by The Open University, using the Open University TeX System.

Printed and bound in the United Kingdom by Hobbs the Printers Limited, Brunel Road, Totton, Hampshire SO40 3WX.

ISBN 0 7492 0225 4

1.1

Contents

Introduction

You have already met a wide range of functions that map the real line to itself. For example, you are familiar with the function t that maps each point x on the real line to the point x^2. This function is denoted by

$$t : \mathbb{R} \longrightarrow \mathbb{R}$$
$$x \longmapsto x^2.$$

There are, however, many important functions where the domain and codomain are not the real line. For example, we can define a function with domain and codomain $\mathbb{R}^2$ that reflects the plane in the x-axis:

For a function $t : V \longrightarrow W$, the domain is V and the codomain is W.

$$t : \mathbb{R}^2 \longrightarrow \mathbb{R}^2$$
$$(x, y) \longmapsto (x, -y).$$

We can also define a function with domain and codomain $\mathbb{R}^3$ that rotates $\mathbb{R}^3$ about the z-axis through an angle $\pi/2$. These are both examples of functions that occur in practical situations.

All the functions mentioned so far are examples of functions between *vector spaces*. In this unit we study functions between general vector spaces.

In Section 1 we look at four important classes of functions acting on the plane $\mathbb{R}^2$: *dilations*, *stretchings*, *rotations* and *reflections*. We find that the functions in these four classes map parallel lines to parallel lines, preserve scalar multiples and map the zero vector to itself. Any function between vector spaces that has all these geometric properties is called a *linear transformation*.

These functions are examples of linear transformations.

The functions acting on the plane that we study in Section 1 all have *matrix representations*. For example, a reflection of $\mathbb{R}^2$ in the x-axis can be represented by

$$\begin{pmatrix} x \\ y \end{pmatrix} \longmapsto \begin{pmatrix} 1 & 0 \\ 0 & -1 \end{pmatrix} \begin{pmatrix} x \\ y \end{pmatrix} = \begin{pmatrix} x \\ -y \end{pmatrix}.$$

We say that

$$\begin{pmatrix} 1 & 0 \\ 0 & -1 \end{pmatrix}$$

is a matrix of the reflection. In Section 2 we show that the linear transformations from a finite-dimensional vector space V to a finite-dimensional vector space W are precisely those functions $t : V \longrightarrow W$ that have matrix representations.

This link between linear transformations and matrices enables us to relate the properties of matrices that we studied in Unit LA2 to the properties of linear transformations. In Section 3 we show that the composite $s \circ t$ of two linear transformations s and t is always a linear transformation, and that a matrix of $s \circ t$ can be found by multiplying a matrix of s and a matrix of t. We also show that a linear transformation t has an inverse when a matrix of t is invertible.

In Section 4 we prove an important result concerning linear transformations, known as the *Dimension Theorem*. This has a number of consequences. For example, it enables us to show how the number of solutions to a system of m simultaneous linear equations in n unknowns depends on the values of m and n.

Study guide

The sections should be read in the natural order. Subsection 2.1 includes the audio section. There are only four sections, but it may take you more than two study sessions to work through Sections 3 and 4.

Before studying this unit, make sure that you understand the main ideas of Units LA2 and LA3. Many results from these units are used here.

1   Introducing linear transformations

After working through this section, you should be able to:

(a) explain what is meant by a *linear transformation*;
(b) recognise simple linear transformations of the plane;
(c) determine whether or not a given function is a linear transformation;
(d) understand that linear transformations preserve the zero vector and linear combinations of vectors.

1.1   What is a linear transformation?

We begin by investigating the properties of some simple but important functions which map the vector space $\mathbb{R}^2$ to itself. In each case, we illustrate the effect of the function on a square with corners at $(0,0)$, $(0,1)$, $(1,1)$ and $(1,0)$, and the effect on the vector $(1,1)$. We shade half the square for clarity.

A k-**dilation** of $\mathbb{R}^2$ stretches (or scales) vectors radially from the origin by a factor k.

Here k is any real number.

When $k = 2$, the length of a vector is doubled.

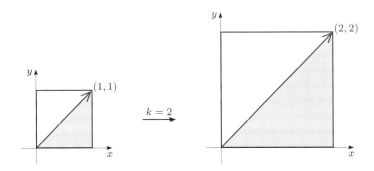

When $k = \frac{1}{2}$, the length of a vector is halved.

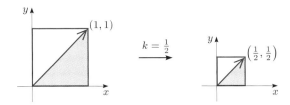

When k is negative, the direction of a vector is reversed —as shown below in the case $k = -2$.

5

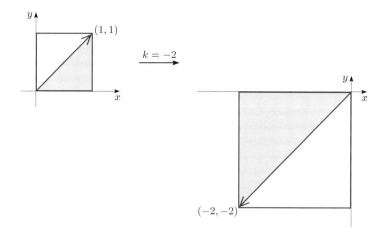

A (k, l)-**stretching** of $\mathbb{R}^2$ stretches (or scales) vectors by a factor k in the x-direction and by a factor l in the y-direction.

Here k and l are any real numbers.

The diagram below shows the effect of a $(2, \frac{1}{2})$-stretching.

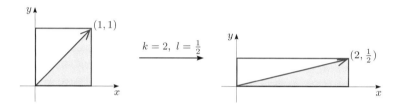

The next diagram shows the effect of a $(-1, 3)$-stretching.

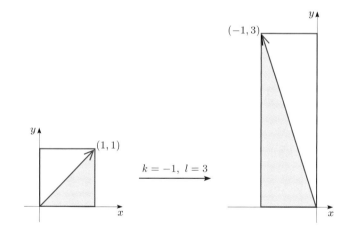

A **rotation** r_θ of $\mathbb{R}^2$ rotates vectors anticlockwise through an angle θ about the origin $(0, 0)$.

The diagram below shows the effect of a rotation r_θ when $\theta = \pi/4$.

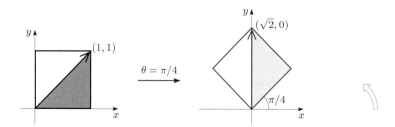

The next diagram shows the effect of a rotation r_θ when $\theta = \pi/2$.

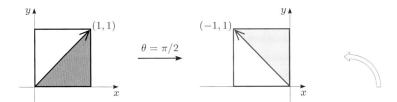

A **reflection** q_ϕ of $\mathbb{R}^2$ reflects vectors in the straight line through the origin that makes an angle ϕ with the x-axis.

The diagram below shows the effect of a reflection q_ϕ when $\phi = \pi/4$.

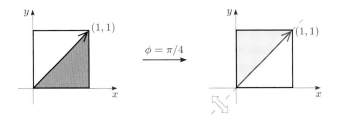

The next diagram shows the effect of a reflection q_ϕ when $\phi = \pi/2$.

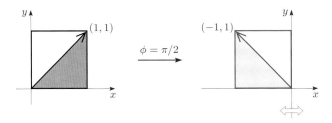

Exercise 1.1   For each of the following functions, draw a diagram to show the effect of the function on the rectangle with corners at $(0,0)$, $(2,0)$, $(2,1)$ and $(0,1)$, and on the vector $(2,1)$. State whether the function is a dilation, a stretching, a rotation or a reflection.

(a) $t : \mathbb{R}^2 \longrightarrow \mathbb{R}^2$
  $(x, y) \longmapsto (2x, 3y)$

(b) $t : \mathbb{R}^2 \longrightarrow \mathbb{R}^2$
  $(x, y) \longmapsto (x, -y)$

(c) $t : \mathbb{R}^2 \longrightarrow \mathbb{R}^2$
  $(x, y) \longmapsto (-y, x)$

We now obtain algebraic definitions of the four types of function defined geometrically above.

A k-dilation of $\mathbb{R}^2$ maps (x, y) to (kx, ky). This can be represented by

$$\begin{pmatrix} x \\ y \end{pmatrix} \longmapsto \begin{pmatrix} k & 0 \\ 0 & k \end{pmatrix} \begin{pmatrix} x \\ y \end{pmatrix} = \begin{pmatrix} kx \\ ky \end{pmatrix}.$$

A (k, l)-stretching of $\mathbb{R}^2$ maps (x, y) to (kx, ly). This can be represented by

$$\begin{pmatrix} x \\ y \end{pmatrix} \longmapsto \begin{pmatrix} k & 0 \\ 0 & l \end{pmatrix} \begin{pmatrix} x \\ y \end{pmatrix} = \begin{pmatrix} kx \\ ly \end{pmatrix}.$$

An algebraic definition for a rotation r_θ of $\mathbb{R}^2$ can be obtained by considering the following diagram, where r_θ maps (x, y) to (x', y').

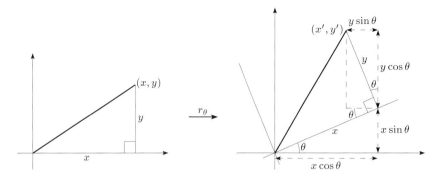

It can be seen that

$$(x, y) \longmapsto (x', y') = (x \cos \theta - y \sin \theta, x \sin \theta + y \cos \theta).$$

This can be represented by

$$\begin{pmatrix} x \\ y \end{pmatrix} \longmapsto \begin{pmatrix} \cos \theta & -\sin \theta \\ \sin \theta & \cos \theta \end{pmatrix} \begin{pmatrix} x \\ y \end{pmatrix} = \begin{pmatrix} x \cos \theta - y \sin \theta \\ x \sin \theta + y \cos \theta \end{pmatrix}.$$

For example, $r_{\pi/6}$ can be represented by

$$\begin{pmatrix} x \\ y \end{pmatrix} \longmapsto \begin{pmatrix} \frac{\sqrt{3}}{2} & -\frac{1}{2} \\ \frac{1}{2} & \frac{\sqrt{3}}{2} \end{pmatrix} \begin{pmatrix} x \\ y \end{pmatrix} = \begin{pmatrix} \frac{\sqrt{3}}{2}x - \frac{1}{2}y \\ \frac{1}{2}x + \frac{\sqrt{3}}{2}y \end{pmatrix}.$$

Similarly, it can be shown that a reflection q_ϕ of $\mathbb{R}^2$ can be defined algebraically by

> We omit the details here, but see the solution to Exercise 1.7.

$$(x, y) \longmapsto (x \cos 2\phi + y \sin 2\phi, x \sin 2\phi - y \cos 2\phi).$$

This can be represented by

$$\begin{pmatrix} x \\ y \end{pmatrix} \longmapsto \begin{pmatrix} \cos 2\phi & \sin 2\phi \\ \sin 2\phi & -\cos 2\phi \end{pmatrix} \begin{pmatrix} x \\ y \end{pmatrix} = \begin{pmatrix} x \cos 2\phi + y \sin 2\phi \\ x \sin 2\phi - y \cos 2\phi \end{pmatrix}.$$

For example, $q_{\pi/6}$ can be represented by

$$\begin{pmatrix} x \\ y \end{pmatrix} \longmapsto \begin{pmatrix} \frac{1}{2} & \frac{\sqrt{3}}{2} \\ \frac{\sqrt{3}}{2} & -\frac{1}{2} \end{pmatrix} \begin{pmatrix} x \\ y \end{pmatrix} = \begin{pmatrix} \frac{1}{2}x + \frac{\sqrt{3}}{2}y \\ \frac{\sqrt{3}}{2} - \frac{1}{2}y \end{pmatrix}.$$

We have seen that each of the four types of function can be represented by

$$\begin{pmatrix} x \\ y \end{pmatrix} \longmapsto \begin{pmatrix} a & b \\ c & d \end{pmatrix} \begin{pmatrix} x \\ y \end{pmatrix} = \begin{pmatrix} ax + by \\ cx + dy \end{pmatrix}$$

for some real numbers a, b, c and d.

The existence of a matrix representation is not the only property shared by these functions of the plane—they also have several striking geometric properties in common. It is easy to see that each function maps straight lines to straight lines—indeed, each maps parallel lines to parallel lines, so parallelograms are mapped to parallelograms. Each function also maps the origin to itself, so we have the following diagram showing the effect of a general transformation t on two vectors, $\mathbf{v}_1$ and $\mathbf{v}_2$.

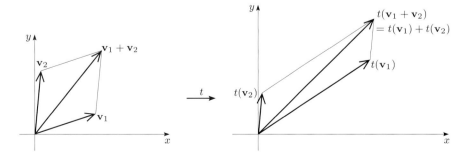

Bearing in mind the Parallelogram Law for vector addition, this illustrates   See Unit LA1, Subsection 2.1.
that for each function t in one of the four classes above,

$$t(\mathbf{v}_1 + \mathbf{v}_2) = t(\mathbf{v}_1) + t(\mathbf{v}_2), \quad \text{for all } \mathbf{v}_1, \mathbf{v}_2 \in \mathbb{R}^2.$$

Such a function t also preserves scalar multiples, as shown below.

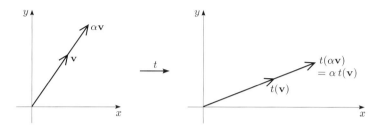

This illustrates that

$$t(\alpha\mathbf{v}) = \alpha\, t(\mathbf{v}), \quad \text{for all } \mathbf{v} \in \mathbb{R}^2,\ \alpha \in \mathbb{R}.$$

We define a linear transformation to be any function from a vector space V The reason for calling such
to a vector space W which has these two algebraic properties.   transformations 'linear' will be
given on page 15.

Definition   Let V and W be vector spaces. A function $t : V \longrightarrow W$ is
a **linear transformation** if it satisfies the following properties.

LT1 $t(\mathbf{v}_1 + \mathbf{v}_2) = t(\mathbf{v}_1) + t(\mathbf{v}_2)$, for all $\mathbf{v}_1, \mathbf{v}_2 \in V$.

LT2 $t(\alpha\mathbf{v}) = \alpha\, t(\mathbf{v})$, for all $\mathbf{v} \in V,\ \alpha \in \mathbb{R}$.

In Section 2 we show that the
functions between
finite-dimensional vector spaces
that have these two properties
are precisely those functions that
have matrix representations.

Suppose that $t : V \longrightarrow W$ is a linear transformation. It follows from   Throughout this unit, V and W
property LT1 that if we know the images of two vectors $\mathbf{v}_1$ and $\mathbf{v}_2$ under t,   denote vector spaces.
then we can find the image of the vector $\mathbf{v}_1 + \mathbf{v}_2$. It follows from property
LT2 that if we know the image of a vector $\mathbf{v}$ under t, then we can find the
image of any scalar multiple of $\mathbf{v}$.

Thus, once we know the images of some vectors, we can find the images of
more vectors by applying properties LT1 and LT2. In fact, if we know the
image of each vector in a *basis* for V, then we can find the image of *every*   We prove this statement at the
vector in V. It is this property that makes linear transformations so   end of this section.
important.

We observed earlier that all the functions of the plane that we have studied
map the origin to itself. In fact, any linear transformation $t : V \longrightarrow W$
maps the zero vector of V to the zero vector of W. To see this, note that

$$t(\mathbf{0}) = t(0\mathbf{0}) = 0t(\mathbf{0}) = \mathbf{0}.$$

We have proved the following result.

Theorem 1.1   Let $t : V \longrightarrow W$ be a linear transformation. Then
$t(\mathbf{0}) = \mathbf{0}$.

It follows from Theorem 1.1 that a function $t : V \longrightarrow W$ where $t(\mathbf{0}) \neq \mathbf{0}$ is
not a linear transformation; for example, the function

$$t : \mathbb{R}^2 \longrightarrow \mathbb{R}^2$$

$$(x, y) \longmapsto (y - 1, x)$$

is not a linear transformation because

$$t(\mathbf{0}) = t(0, 0) = (-1, 0) \neq \mathbf{0}.$$

We include this test in the following strategy.

Strategy 1.1 To determine whether or not a given function
$t : V \longrightarrow W$ is a linear transformation.

1. Check whether $t(\mathbf{0}) = \mathbf{0}$; if not, then t is not a linear
    transformation.

2. Check whether t satisfies the following.

    LT1 $t(\mathbf{v}_1 + \mathbf{v}_2) = t(\mathbf{v}_1) + t(\mathbf{v}_2)$, for all $\mathbf{v}_1, \mathbf{v}_2 \in V$.

    LT2 $t(\alpha \mathbf{v}) = \alpha\, t(\mathbf{v})$, for all $\mathbf{v} \in V$, $\alpha \in \mathbb{R}$.

    The function t is a linear transformation if and only if both these
    properties are satisfied.

A function t with the property
$t(\mathbf{0}) = \mathbf{0}$ is not necessarily a
linear transformation.

The function $t : \mathbb{R}^2 \longrightarrow \mathbb{R}^2$
$(x, y) \longmapsto (x, |y|)$ satisfies
$t(\mathbf{0}) = \mathbf{0}$ but is not a linear
transformation.

If either of LT1 or LT2 fails, then you do not need to check the other.

Example 1.1   Use Strategy 1.1 to determine whether or not each of the
following functions is a linear transformation.

(a) $t : \mathbb{R}^2 \longrightarrow \mathbb{R}^2$
     (b) $t : \mathbb{R}^2 \longrightarrow \mathbb{R}^2$

    $(x, y) \longmapsto (2x, y)$
        $(x, y) \longmapsto ((x + y)^2, y^2)$

Solution

(a) Here $t(\mathbf{0}) = \mathbf{0}$, so t may be a linear transformation.

You may have noticed that t is a
$(2, 1)$-stretching, so we expect it
to be a linear transformation.

Next we check whether t satisfies LT1:

$$t(\mathbf{v}_1 + \mathbf{v}_2) = t(\mathbf{v}_1) + t(\mathbf{v}_2), \quad \text{for all } \mathbf{v}_1, \mathbf{v}_2 \in \mathbb{R}^2.$$

In $\mathbb{R}^2$, let $\mathbf{v}_1 = (x_1, y_1)$ and $\mathbf{v}_2 = (x_2, y_2)$. Then

$$\begin{aligned}
t(\mathbf{v}_1 + \mathbf{v}_2) &= t(x_1 + x_2, y_1 + y_2) \\
&= (2(x_1 + x_2), y_1 + y_2) \\
&= (2x_1 + 2x_2, y_1 + y_2)
\end{aligned}$$

and

$$\begin{aligned}
t(\mathbf{v}_1) + t(\mathbf{v}_2) &= (2x_1, y_1) + (2x_2, y_2) \\
&= (2x_1 + 2x_2, y_1 + y_2).
\end{aligned}$$

These expressions are equal, so LT1 is satisfied.

Finally, we check whether t satisfies LT2:

$$t(\alpha \mathbf{v}) = \alpha\, t(\mathbf{v}), \quad \text{for all } \mathbf{v} \in \mathbb{R}^2, \ \alpha \in \mathbb{R}.$$

Let $\mathbf{v} = (x, y)$ be a vector in $\mathbb{R}^2$ and let $\alpha \in \mathbb{R}$. Then

$$t(\alpha \mathbf{v}) = t(\alpha x, \alpha y) = (2\alpha x, \alpha y)$$

and

$$\alpha\, t(\mathbf{v}) = \alpha\, t(x, y) = \alpha(2x, y) = (2\alpha x, \alpha y).$$

These expressions are equal, so LT2 is satisfied.

Since LT1 and LT2 are satisfied, t is a linear transformation.

(b) Here $t(\mathbf{0}) = \mathbf{0}$, so t may be a linear transformation.

Next we check whether t satisfies LT1:

$$t(\mathbf{v}_1 + \mathbf{v}_2) = t(\mathbf{v}_1) + t(\mathbf{v}_2), \quad \text{for all } \mathbf{v}_1, \mathbf{v}_2 \in \mathbb{R}^2.$$

In $\mathbb{R}^2$, let $\mathbf{v}_1 = (x_1, y_1)$ and $\mathbf{v}_2 = (x_2, y_2)$. Then

$$t(\mathbf{v}_1 + \mathbf{v}_2) = t(x_1 + x_2, y_1 + y_2)$$
$$= ((x_1 + x_2 + y_1 + y_2)^2, (y_1 + y_2)^2)$$

and

$$t(\mathbf{v}_1) + t(\mathbf{v}_2) = ((x_1 + y_1)^2, y_1^2) + ((x_2 + y_2)^2, y_2^2)$$
$$= ((x_1 + y_1)^2 + (x_2 + y_2)^2, y_1^2 + y_2^2).$$

These expressions are not equal in general, so LT1 is not satisfied.

Thus t is not a linear transformation.   ■

Since LT1 is not satisfied, we do not need to check LT2.

Exercise 1.2   Determine whether or not each of the following functions is a linear transformation.

(a) $t : \mathbb{R}^2 \longrightarrow \mathbb{R}^2$
$\qquad (x, y) \longmapsto (x + 3y, y)$

(b) $t : \mathbb{R}^2 \longrightarrow \mathbb{R}^2$
$\qquad (x, y) \longmapsto (x + 2, y + 1)$

In Exercise 1.2(a), you showed that the function

$t : \mathbb{R}^2 \longrightarrow \mathbb{R}^2$

$(x, y) \longmapsto (x + 3y, y)$

is a linear transformation. This function is an example of a *shear*, or *skew*, of $\mathbb{R}^2$.

In general, a **shear** of $\mathbb{R}^2$ in the x-direction by a factor k is the linear transformation

$t : \mathbb{R}^2 \longrightarrow \mathbb{R}^2$

$(x, y) \longmapsto (x + ky, y).$

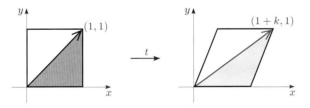

In Exercise 1.2(b), you showed that the function

$t : \mathbb{R}^2 \longrightarrow \mathbb{R}^2$

$(x, y) \longmapsto (x + 2, y + 1)$

is not a linear transformation. This function is an example of a *translation* of $\mathbb{R}^2$.

In general, a **translation** of $\mathbb{R}^2$ by (a, b) is the function

$t : \mathbb{R}^2 \longrightarrow \mathbb{R}^2$

$(x, y) \longmapsto (x + a, y + b).$

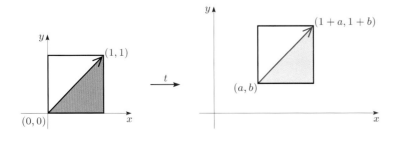

11

A translation is not a linear transformation unless $a = b = 0$, since otherwise it does not map the origin to itself.

1.2 Examples of linear transformations

We have seen many examples of functions from $\mathbb{R}^2$ to $\mathbb{R}^2$. In general, given any two vector spaces V and W, we can define functions from V to W.

Example 1.2 We can define a function t from $\mathbb{R}^3$ to $\mathbb{R}^2$ by projecting each vector in $\mathbb{R}^3$ onto the (x, y)-plane:

$$t : \mathbb{R}^3 \longrightarrow \mathbb{R}^2$$
$$(x, y, z) \longmapsto (x, y).$$

Show that this function is a linear transformation.

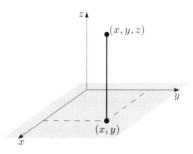

Solution First we show that t satisfies LT1:

$$t(\mathbf{v}_1 + \mathbf{v}_2) = t(\mathbf{v}_1) + t(\mathbf{v}_2), \quad \text{for all } \mathbf{v}_1, \mathbf{v}_2 \in \mathbb{R}^3.$$

In $\mathbb{R}^3$, let $\mathbf{v}_1 = (x_1, y_1, z_1)$ and $\mathbf{v}_2 = (x_2, y_2, z_2)$. Then

$$t(\mathbf{v}_1 + \mathbf{v}_2) = t(x_1 + x_2, y_1 + y_2, z_1 + z_2)$$
$$= (x_1 + x_2, y_1 + y_2)$$

and

$$t(\mathbf{v}_1) + t(\mathbf{v}_2) = t(x_1, y_1, z_1) + t(x_2, y_2, z_2)$$
$$= (x_1, y_1) + (x_2, y_2)$$
$$= (x_1 + x_2, y_1 + y_2).$$

These expressions are equal, so LT1 is satisfied.

Next we show that t satisfies LT2:

$$t(\alpha \mathbf{v}) = \alpha \, t(\mathbf{v}), \quad \text{for all } \mathbf{v} \in \mathbb{R}^3, \ \alpha \in \mathbb{R}.$$

Let $\mathbf{v} = (x, y, z)$ be a vector in $\mathbb{R}^3$ and let $\alpha \in \mathbb{R}$. Then

$$t(\alpha \mathbf{v}) = t(\alpha x, \alpha y, \alpha z) = (\alpha x, \alpha y)$$

and

$$\alpha \, t(\mathbf{v}) = \alpha \, t(x, y, z) = \alpha(x, y) = (\alpha x, \alpha y).$$

These expressions are equal, so LT2 is satisfied.

Since LT1 and LT2 are satisfied, t is a linear transformation. ■

Here, we use the definition (on page 9) rather than Strategy 1.1, as we are asked to 'show', not 'determine'.

Example 1.3 Determine whether or not the following function is a linear transformation.

$$t : \mathbb{R}^4 \longrightarrow \mathbb{R}^2$$
$$(x, y, z, w) \longmapsto (xy, z)$$

Solution We use Strategy 1.1.

Since $t(\mathbf{0}) = \mathbf{0}$, t may be a linear transformation.

Next we check whether t satisfies LT1:

$$t(\mathbf{v}_1 + \mathbf{v}_2) = t(\mathbf{v}_1) + t(\mathbf{v}_2), \quad \text{for all } \mathbf{v}_1, \mathbf{v}_2 \in \mathbb{R}^4.$$

In $\mathbb{R}^4$, let $\mathbf{v}_1 = (x_1, y_1, z_1, w_1)$ and $\mathbf{v}_2 = (x_2, y_2, z_2, w_2)$. Then

$$t(\mathbf{v}_1 + \mathbf{v}_2) = t(x_1 + x_2, y_1 + y_2, z_1 + z_2, w_1 + w_2)$$
$$= ((x_1 + x_2)(y_1 + y_2), z_1 + z_2)$$

and

$$t(\mathbf{v}_1) + t(\mathbf{v}_2) = (x_1 y_1, z_1) + (x_2 y_2, z_2)$$
$$= (x_1 y_1 + x_2 y_2, z_1 + z_2).$$

Since $(x_1 + x_2)(y_1 + y_2) \neq x_1 y_1 + x_2 y_2$ in general, LT1 is not satisfied.

Thus t is not a linear transformation.   ■

Exercise 1.3   Determine whether or not each of the following functions is a linear transformation.

(a) $t : \mathbb{R}^2 \longrightarrow \mathbb{R}^4$
$$(x, y) \longmapsto (x, y, x, y)$$

(b) $t : \mathbb{R}^3 \longrightarrow \mathbb{R}$
$$(x, y, z) \longmapsto x^2$$

(c) $t : \mathbb{R}^3 \longrightarrow \mathbb{R}^4$
$$(x, y, z) \longmapsto (x, y, z, 1)$$

In the previous subsection we gave an algebraic definition of a rotation of $\mathbb{R}^2$. Similarly, a rotation of $\mathbb{R}^3$ in an anticlockwise direction about the z-axis through an angle θ is given by

$$t : \mathbb{R}^3 \longrightarrow \mathbb{R}^3$$
$$(x, y, z) \longmapsto (x \cos \theta - y \sin \theta, x \sin \theta + y \cos \theta, z).$$

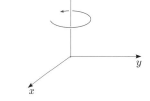

Exercise 1.4   Show that the function t above is a linear transformation.

So far we have considered functions $t : V \longrightarrow W$ where $V = \mathbb{R}^n$ and $W = \mathbb{R}^m$ for some $m, n \in \mathbb{N}$. There are, however, functions between other types of vector space.

For example, let P_3 and P_2 be the vector spaces of polynomials defined by

$$P_3 = \{p(x) : p(x) = a + bx + cx^2, \ a, b, c \in \mathbb{R}\},$$
$$P_2 = \{p(x) : p(x) = a + bx, \ a, b \in \mathbb{R}\}.$$

We introduced these vector spaces in Unit LA3, Subsection 1.2.

Example 1.4   Consider the function which maps each polynomial $p(x)$, where $p(x) = a + bx + cx^2$, in P_3 to its derivative $p'(x) = b + 2cx$ in P_2:

$$t : P_3 \longrightarrow P_2$$
$$p(x) \longmapsto p'(x).$$

Determine whether or not this function is a linear transformation.

Solution   We use Strategy 1.1.

Since the zero element of P_3 is $p(x) = 0$, we have $t(\mathbf{0}) = \mathbf{0}$, so t may be a linear transformation.

Next we check whether t satisfies LT1:

$$t(p(x) + q(x)) = t(p(x)) + t(q(x)), \quad \text{for all } p(x), q(x) \in P_3.$$

Let $p(x), q(x) \in P_3$. Then

$$t(p(x) + q(x)) = (p(x) + q(x))' = p'(x) + q'(x)$$

and

$$t(p(x)) + t(q(x)) = p'(x) + q'(x).$$

These expressions are equal, so LT1 is satisfied.

Finally, we check whether t satisfies LT2:

$$t(\alpha p(x)) = \alpha\, t(p(x)), \quad \text{for all } p(x) \in P_3,\ \alpha \in \mathbb{R}.$$

Let $p(x) \in P_3$ and $\alpha \in \mathbb{R}$. Then

$$t(\alpha p(x)) = (\alpha p(x))' = \alpha p'(x)$$

and

$$\alpha\, t(p(x)) = \alpha p'(x).$$

These expressions are equal, so LT2 is satisfied.

Since LT1 and LT2 are satisfied, t is a linear transformation.   ∎

> **Exercise 1.5**   Consider the function t from P_3 to itself obtained by adding to each polynomial $p(x) = a + bx + cx^2$ in P_3 the number $p(2) = a + 2b + 4c$:
>
> $$t : P_3 \longrightarrow P_3$$
> $$p(x) \longmapsto p(x) + p(2).$$
>
> Determine whether or not this function is a linear transformation.

There are also linear transformations of infinite-dimensional vector spaces. For example, let V be the vector space of all real functions. An argument similar to that in the solution to Exercise 1.5 shows that the following function is a linear transformation:

$$t : V \longrightarrow V$$
$$f(x) \longmapsto f(x) + f(2).$$

Zero transformation

Given any two vector spaces V and W, we can define a particularly simple function by mapping each vector in V to the zero vector in W:

$$t : V \longrightarrow W$$
$$\mathbf{v} \longmapsto \mathbf{0}.$$

We now show that this function is a linear transformation.

First we show that t satisfies LT1:

$$t(\mathbf{v}_1 + \mathbf{v}_2) = t(\mathbf{v}_1) + t(\mathbf{v}_2), \quad \text{for all } \mathbf{v}_1, \mathbf{v}_2 \in V.$$

Let $\mathbf{v}_1, \mathbf{v}_2 \in V$. Then $\mathbf{v}_1 + \mathbf{v}_2$ is also in V, so

$$t(\mathbf{v}_1 + \mathbf{v}_2) = \mathbf{0}$$

and

$$t(\mathbf{v}_1) + t(\mathbf{v}_2) = \mathbf{0} + \mathbf{0} = \mathbf{0}.$$

So LT1 is satisfied.

Next we show that t satisfies LT2:

$$t(\alpha \mathbf{v}) = \alpha\, t(\mathbf{v}), \quad \text{for all } \mathbf{v} \in V,\ \alpha \in \mathbb{R}.$$

Let $\mathbf{v} \in V$ and $\alpha \in \mathbb{R}$. Then

$$t(\alpha \mathbf{v}) = \mathbf{0}$$

and

$$\alpha\, t(\mathbf{v}) = \alpha \mathbf{0} = \mathbf{0}.$$

Every vector space contains a zero vector, and a linear transformation maps the zero vector in its domain to the zero vector in its codomain.

So LT2 is satisfied.

Since LT1 and LT2 are satisfied, t is a linear transformation. We give it a special name.

Definition   The **zero transformation** from V to W is the linear transformation

$$t : V \longrightarrow W$$
$$\mathbf{v} \longmapsto \mathbf{0}.$$

Identity transformation

Given a vector space V, we can define a function from V to itself by mapping each vector in V to itself:

$$i_V : V \longrightarrow V$$
$$\mathbf{v} \longmapsto \mathbf{v}.$$

Exercise 1.6   Show that the function i_V is a linear transformation.

Definition   The **identity transformation** of V is the linear transformation

$$i_V : V \longrightarrow V$$
$$\mathbf{v} \longmapsto \mathbf{v}.$$

We omit the subscript V when the vector space is clear from the context.

1.3   Linear combinations of vectors

We end this section by proving that linear combinations of vectors are preserved under a linear transformation. This explains why these functions are called linear transformations.

A linear combination of the vectors $\mathbf{v}_1, \ldots, \mathbf{v}_n$ is an expression of the form $\alpha_1 \mathbf{v}_1 + \cdots + \alpha_n \mathbf{v}_n$, where $\alpha_1, \ldots, \alpha_n \in \mathbb{R}$. (See Unit LA3, Subsection 2.1.)

Some texts use this as the definition of a linear transformation.

Theorem 1.2   A function $t : V \longrightarrow W$ is a linear transformation if and only if it satisfies

LT3 $t(\alpha_1 \mathbf{v}_1 + \alpha_2 \mathbf{v}_2) = \alpha_1 t(\mathbf{v}_1) + \alpha_2 t(\mathbf{v}_2)$,
    for all $\mathbf{v}_1, \mathbf{v}_2 \in V$ and all $\alpha_1, \alpha_2 \in \mathbb{R}$.

Proof   We begin by using LT1 and LT2 to show that if a function $t : V \longrightarrow W$ is a linear transformation, then it satisfies LT3.

Let $\mathbf{v}_1, \mathbf{v}_2 \in V$ and $\alpha_1, \alpha_2 \in \mathbb{R}$. Then it follows from LT1 that

$$t(\alpha_1 \mathbf{v}_1 + \alpha_2 \mathbf{v}_2) = t(\alpha_1 \mathbf{v}_1) + t(\alpha_2 \mathbf{v}_2),$$

and from LT2 that

$$t(\alpha_1 \mathbf{v}_1) + t(\alpha_2 \mathbf{v}_2) = \alpha_1 t(\mathbf{v}_1) + \alpha_2 t(\mathbf{v}_2).$$

So t satisfies the property LT3:

$$t(\alpha_1 \mathbf{v}_1 + \alpha_2 \mathbf{v}_2) = \alpha_1 t(\mathbf{v}_1) + \alpha_2 t(\mathbf{v}_2),$$
    for all $\mathbf{v}_1, \mathbf{v}_2 \in V$ and all $\alpha_1, \alpha_2 \in \mathbb{R}$.

Now we prove the converse. Suppose that a function $t : V \longrightarrow W$ satisfies property LT3. Then it also satisfies LT1 and LT2, since

$$t(\mathbf{v}_1 + \mathbf{v}_2) = t(\mathbf{v}_1) + t(\mathbf{v}_2), \quad \text{for all } \mathbf{v}_1, \mathbf{v}_2 \in V,$$

is a special case of LT3 with $\alpha_1 = \alpha_2 = 1$, and

$$t(\alpha \mathbf{v}) = \alpha\, t(\mathbf{v}), \quad \text{for all } \mathbf{v} \in V,\ \alpha \in \mathbb{R},$$

is a special case of LT3 with $\mathbf{v}_2 = \mathbf{0}$, $\mathbf{v}_1 = \mathbf{v}$ and $\alpha_1 = \alpha$.

Thus a function is a linear transformation if and only if it satisfies property LT3. ∎

We now prove that linear combinations of any number of vectors are preserved under a linear transformation.

Theorem 1.3   Let $t : V \longrightarrow W$ be a linear transformation. Then

$$t(\alpha_1 \mathbf{v}_1 + \alpha_2 \mathbf{v}_2 + \cdots + \alpha_n \mathbf{v}_n) = \alpha_1 t(\mathbf{v}_1) + \alpha_2 t(\mathbf{v}_2) + \cdots + \alpha_n t(\mathbf{v}_n),$$

for all $\mathbf{v}_1, \ldots, \mathbf{v}_n \in V$ and all $\alpha_1, \ldots, \alpha_n \in \mathbb{R}$, $n \in \mathbb{N}$.

Proof   We use proof by mathematical induction.   Unit I2

Let $P(n)$ be the statement

$$t(\alpha_1 \mathbf{v}_1 + \alpha_2 \mathbf{v}_2 + \cdots + \alpha_n \mathbf{v}_n) = \alpha_1 t(\mathbf{v}_1) + \alpha_2 t(\mathbf{v}_2) + \cdots + \alpha_n t(\mathbf{v}_n),$$
for all $\mathbf{v}_1, \ldots, \mathbf{v}_n \in V$ and all $\alpha_1, \ldots, \alpha_n \in \mathbb{R}$.

STEP 1   Since t is a linear transformation, LT2 is satisfied, so

$$t(\alpha_1 \mathbf{v}_1) = \alpha_1 t(\mathbf{v}_1), \quad \text{for all } \mathbf{v}_1 \in V,\ \alpha_1 \in \mathbb{R}.$$

Thus $P(1)$ is true.

STEP 2   We assume that $P(k)$ is true for some positive integer k; that is,

$$t(\alpha_1 \mathbf{v}_1 + \alpha_2 \mathbf{v}_2 + \cdots + \alpha_k \mathbf{v}_k) = \alpha_1 t(\mathbf{v}_1) + \alpha_2 t(\mathbf{v}_2) + \cdots + \alpha_k t(\mathbf{v}_k),$$
for all $\mathbf{v}_1, \ldots, \mathbf{v}_k \in V$ and all $\alpha_1, \ldots, \alpha_k \in \mathbb{R}$.

Then for all $\mathbf{v}_1, \ldots, \mathbf{v}_{k+1} \in V$ and all $\alpha_1, \ldots, \alpha_{k+1} \in \mathbb{R}$,

$$t(\alpha_1 \mathbf{v}_1 + \alpha_2 \mathbf{v}_2 + \cdots + \alpha_k \mathbf{v}_k + \alpha_{k+1} \mathbf{v}_{k+1})$$
$$= t((\alpha_1 \mathbf{v}_1 + \alpha_2 \mathbf{v}_2 + \cdots + \alpha_k \mathbf{v}_k) + \alpha_{k+1} \mathbf{v}_{k+1})$$
$$= t(\alpha_1 \mathbf{v}_1 + \alpha_2 \mathbf{v}_2 + \cdots + \alpha_k \mathbf{v}_k) + t(\alpha_{k+1} \mathbf{v}_{k+1}) \quad \text{(by LT1)}$$
$$= t(\alpha_1 \mathbf{v}_1 + \alpha_2 \mathbf{v}_2 + \cdots + \alpha_k \mathbf{v}_k) + \alpha_{k+1} t(\mathbf{v}_{k+1}) \quad \text{(by LT2)}$$
$$= \alpha_1 t(\mathbf{v}_1) + \alpha_2 t(\mathbf{v}_2) + \cdots + \alpha_k t(\mathbf{v}_k) + \alpha_{k+1} t(\mathbf{v}_{k+1}) \quad \text{(by } P(k)\text{)}.$$

Thus, for $k = 1, 2, \ldots$, $P(k)$ is true implies that $P(k+1)$ is true.

Hence, by the Principle of Mathematical Induction, $P(n)$ is true for all $n \in \mathbb{N}$. ∎

Theorem 1.3 is an important result. It means that, given a linear transformation $t : V \longrightarrow W$ and the images of each of the vectors in a basis for V, we can determine the image of *any* vector in V.

Consider the linear transformation r_θ which rotates each vector in $\mathbb{R}^2$ anticlockwise through an angle θ about the origin. The standard basis for $\mathbb{R}^2$ is $\{(1,0), (0,1)\}$. From the diagram in the margin we can check that

$$r_\theta(1,0) = (\cos\theta, \sin\theta) \quad \text{and} \quad r_\theta(0,1) = (-\sin\theta, \cos\theta).$$

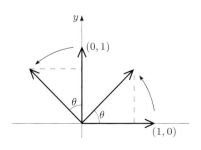

We now write each vector (x, y) in $\mathbb{R}^2$ in the form

$$(x, y) = x(1, 0) + y(0, 1),$$

so, from Theorem 1.3,

$$r_\theta(x, y) = x(\cos\theta, \sin\theta) + y(-\sin\theta, \cos\theta)$$
$$= (x\cos\theta - y\sin\theta, x\sin\theta + y\cos\theta).$$

This method of finding an algebraic definition for r_θ is simpler than the geometric approach used in Subsection 1.1 and is more generally applicable.

> **Exercise 1.7**   Find the image of a vector (x, y) in $\mathbb{R}^2$ under the reflection q_ϕ, given that $q_\phi(1, 0) = (\cos 2\phi, \sin 2\phi)$ and $q_\phi(0, 1) = (\sin 2\phi, -\cos 2\phi)$.

Further exercises

Exercise 1.8   Determine whether or not each of the following functions is a linear transformation.

(a) $t : \mathbb{R}^2 \longrightarrow \mathbb{R}^2$
$(x, y) \longmapsto (3x + y, 2x - y)$

(b) $t : \mathbb{R}^2 \longrightarrow \mathbb{R}^2$
$(x, y) \longmapsto (x + 1, y)$

Exercise 1.9 Determine whether or not each of the following functions is a linear transformation.

(a) $t : \mathbb{R}^2 \longrightarrow \mathbb{R}^3$
$(x, y) \longmapsto (x^2, y, y)$

(b) $t : \mathbb{R}^3 \longmapsto \mathbb{R}^2$
$(x, y, z) \longmapsto (x + y, x - z)$

(c) $t : \mathbb{R}^2 \longrightarrow \mathbb{R}^4$
$(x, y) \longmapsto (x, y, 1, x)$

Exercise 1.10 Determine whether or not each of the following functions is a linear transformation.

(a) $t : P_3 \longrightarrow P_3$
$p(x) \longmapsto p(x) + p'(x)$

(b) $t : P_3 \longrightarrow P_3$
$p(x) \longmapsto p(x) + 2$

Exercise 1.11 Let V be the vector space

$$V = \{f(x) : f(x) = ae^x \cos x + be^x \sin x, \ a, b \in \mathbb{R}\}.$$

Determine whether or not each of the following functions is a linear transformation.

(a) $t : V \longrightarrow V$
$f(x) \longmapsto f'(x)$

(b) $t : V \longrightarrow \mathbb{R}^2$
$ae^x \cos x + be^x \sin x \longmapsto (2a, b + 1)$

2   Matrices of linear transformations

After working through this section, you should be able to:

(a) find the *matrix representation* of a linear transformation $t : V \longrightarrow W$ with respect to given bases for the finite-dimensional vector spaces V and W;

(b) understand that the matrix representation of a linear transformation $t : V \longrightarrow W$ depends on the bases used for V and W;

(c) understand the relationship between matrices and linear transformations.

2.1   Finding matrix representations

In Section 1 you met several examples of matrix representations of linear transformations. For example, you saw that a 2-dilation of $\mathbb{R}^2$ can be represented by

$$\begin{pmatrix} x \\ y \end{pmatrix} \longmapsto \begin{pmatrix} 2 & 0 \\ 0 & 2 \end{pmatrix} \begin{pmatrix} x \\ y \end{pmatrix} = \begin{pmatrix} 2x \\ 2y \end{pmatrix}$$

and a rotation r_θ of $\mathbb{R}^2$ can be represented by

$$\begin{pmatrix} x \\ y \end{pmatrix} \longmapsto \begin{pmatrix} \cos\theta & -\sin\theta \\ \sin\theta & \cos\theta \end{pmatrix} \begin{pmatrix} x \\ y \end{pmatrix} = \begin{pmatrix} x\cos\theta - y\sin\theta \\ x\sin\theta + y\cos\theta \end{pmatrix}.$$

In this section we show that any linear transformation $t : V \longrightarrow W$ between finite-dimensional vector spaces has a matrix representation

$$\begin{pmatrix} v_1 \\ \vdots \\ v_n \end{pmatrix} \longmapsto \begin{pmatrix} a_{11} & a_{12} & \cdots & a_{1n} \\ \vdots & \vdots & & \vdots \\ a_{m1} & a_{m2} & \cdots & a_{mn} \end{pmatrix} \begin{pmatrix} v_1 \\ \vdots \\ v_n \end{pmatrix} = \begin{pmatrix} w_1 \\ \vdots \\ w_m \end{pmatrix},$$

or

$$\mathbf{v}^T \longmapsto \mathbf{A}\mathbf{v}^T = \mathbf{w}^T.$$

Matrix representations are important because they are an aid to performing calculations with linear transformations; in particular, they are easily handled by computers.

We have seen that it is sometimes convenient to use a non-standard basis $E = \{\mathbf{e}_1, \ldots, \mathbf{e}_n\}$ for a vector space V. Recall that if $\mathbf{v}$ is a vector in V and

$$\mathbf{v} = v_1\mathbf{e}_1 + \cdots + v_n\mathbf{e}_n,$$

then the numbers $v_1, \ldots, v_n$ are the *coordinates* of $\mathbf{v}$ with respect to the basis E (the *E-coordinates* of $\mathbf{v}$). The *E-coordinate representation* of $\mathbf{v}$ is $\mathbf{v}_E = (v_1, \ldots, v_n)_E$.

Unit LA3, Subsection 3.3.

For example, let E be the basis $\{(1, 1), (1, 0)\}$ for $\mathbb{R}^2$. The vector $\mathbf{v} = (5, 2)$ in $\mathbb{R}^2$ can be written as

$$\mathbf{v} = 2(1, 1) + 3(1, 0),$$

so the *E-coordinate* representation of $\mathbf{v}$ is

$$\mathbf{v}_E = (2, 3)_E.$$

For another example, consider the basis $E = \{1 + x^2, x^2, 2 - x\}$ for the vector space P_3. As

$$1 + x + 2x^2 = 3(1 + x^2) - x^2 - (2 - x),$$

the E-coordinate representation of the polynomial $1 + x + 2x^2$ is $(3, -1, -1)_E$.

The following exercises are designed to remind you how to write a vector in terms of its coordinates with respect to a basis.

Exercise 2.1   Find the E-coordinate representation of the vector $\mathbf{v} = (3, 1)$ in $\mathbb{R}^2$ for each of the following bases E for $\mathbb{R}^2$.

(a) $E = \{(3, 1), (2, 1)\}$ (b) $E = \{(1, 2), (2, 1)\}$

Exercise 2.2   Find the E-coordinate representation of the polynomial $p(x) = 2 + 3x$ in P_2 for each of the following bases E for P_2.

(a) $E = \{1, x\}$ (the standard basis)

(b) $E = \{1, 4 + 6x\}$

(c) $E = \{2x, 1 + 4x\}$

We now define a *matrix representation* of a linear transformation.

Definitions   Let V and W be vector spaces of dimensions n and m, respectively. Let $t : V \longrightarrow W$ be a linear transformation, let $E = \{\mathbf{e}_1, \ldots, \mathbf{e}_n\}$ be a basis for V, let $F = \{\mathbf{f}_1, \ldots, \mathbf{f}_m\}$ be a basis for W and let $\mathbf{A}$ be an $m \times n$ matrix such that

$$t(\mathbf{v})_F = \mathbf{A}\mathbf{v}_E, \quad \text{for each vector } \mathbf{v} \text{ in } V.$$

Then $\mathbf{v}_E \longmapsto \mathbf{A}\mathbf{v}_E = t(\mathbf{v})_F$ is the **matrix representation** of t with respect to the bases E and F, and $\mathbf{A}$ is the **matrix** of t with respect to the bases E and F.

Later in this section, we *prove* that there is exactly one matrix of t with respect to the bases E and F.

Remarks

1.   A matrix of a linear transformation from an n-dimensional vector space to an m-dimensional vector space is an $m \times n$ matrix, not an $n \times m$ matrix as you might expect.

2.   Strictly speaking, we should write $\mathbf{v}_E^T \longmapsto \mathbf{A}\mathbf{v}_E^T = t(\mathbf{v})_F^T$. We omit the transpose symbols for simplicity.

3.   When $E = F$, we refer to the matrix representation with respect to the basis E.

4.   These definitions make sense only when V and W are finite dimensional.

In the audio section we develop a strategy (Strategy 2.1) for finding matrix representations of linear transformations, and then work through several examples showing how the strategy is used.

Before starting the audio, try the following exercise.

Exercise 2.3   Each of the following linear transformations $t : \mathbb{R}^2 \longrightarrow \mathbb{R}^2$ is defined by a matrix representation with respect to the standard basis $\{(1,0), (0,1)\}$ for $\mathbb{R}^2$. In each case, find the images of the vectors $(1,0)$ and $(0,1)$. What do you notice about the relationship between the vectors $t(1,0)$ and $t(0,1)$ and the 2×2 matrix of the linear transformation?

(a) $t : \mathbb{R}^2 \longrightarrow \mathbb{R}^2$

$$\begin{pmatrix} x \\ y \end{pmatrix} \longmapsto \begin{pmatrix} 3 & 0 \\ 0 & 2 \end{pmatrix} \begin{pmatrix} x \\ y \end{pmatrix}$$

In part (a), t is a $(3,2)$-stretching of $\mathbb{R}^2$.

(b) $t : \mathbb{R}^2 \longrightarrow \mathbb{R}^2$

$$\begin{pmatrix} x \\ y \end{pmatrix} \longmapsto \begin{pmatrix} \frac{1}{\sqrt{2}} & -\frac{1}{\sqrt{2}} \\ \frac{1}{\sqrt{2}} & \frac{1}{\sqrt{2}} \end{pmatrix} \begin{pmatrix} x \\ y \end{pmatrix}$$

In part (b), t is a rotation $r_{\pi/4}$ of $\mathbb{R}^2$.

Listen to the audio as you work through the frames.

Audio

3. Strategy 2.1   Finding the matrix of t

$$\begin{array}{ccc}
\boxed{\begin{array}{c} V \\ \text{basis} \\ E = \{e_1, \ldots, e_n\} \end{array}} & \xrightarrow{\ t\ } & \boxed{\begin{array}{c} W \\ \text{basis} \\ F = \{f_1, \ldots, f_m\} \end{array}}
\end{array}$$

linear transformation

To find the matrix A of t with respect to the bases E and F.

1. Find $t(e_1), \ldots, t(e_n)$.
2. Find the F-coordinates of each of these image vectors.

$$t(e_1) = (a_{11}, a_{21}, \ldots, a_{m1})_F$$
$$t(e_2) = (a_{12}, a_{22}, \ldots, a_{m2})_F$$
$$\cdots$$
$$t(e_n) = (a_{1n}, a_{2n}, \ldots, a_{mn})_F$$

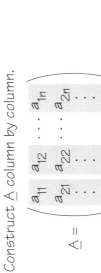

(vectors form columns)

3. Construct A column by column.

$$A = \begin{pmatrix} a_{11} & a_{12} & \cdots & a_{1n} \\ a_{21} & a_{22} & \cdots & a_{2n} \\ \vdots & \vdots & & \vdots \\ a_{m1} & a_{m2} & \cdots & a_{mn} \end{pmatrix}$$

The matrix representation of t with respect to the bases E and F is

$$\begin{pmatrix} v_1 \\ v_2 \\ \cdots \\ v_n \end{pmatrix}_E \longrightarrow \underbrace{\begin{pmatrix} a_{11} & a_{12} & \cdots & a_{1n} \\ a_{21} & a_{22} & \cdots & a_{2n} \\ & & & \\ a_{m1} & a_{m2} & \cdots & a_{mn} \end{pmatrix}}_{A} \longrightarrow \begin{pmatrix} w_1 \\ w_2 \\ \cdots \\ w_m \end{pmatrix}_F$$

$$v_E = \underset{A}{} \quad = w_F$$

---

1. Example 2.1

$$\begin{array}{ccc}
\boxed{\begin{array}{c} \mathbb{R}^3 \\ \text{standard basis} \\ E = \{(1,0,0),(0,1,0),(0,0,1)\} \end{array}} & \xrightarrow{\ t\ } & \boxed{\begin{array}{c} \mathbb{R}^2 \\ \text{standard basis} \\ F = \{(1,0),(0,1)\} \end{array}}
\end{array}$$

$$t(x,y,z) = (x,y)$$

$$t(1,0,0) = (1,0), \ t(0,1,0) = (0,1), \ t(0,0,1) = (0,0).$$

The matrix of t with respect to the standard bases is

$$A = \begin{pmatrix} 1 & 0 & 0 \\ 0 & 1 & 0 \end{pmatrix}.$$

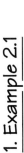

Check: $\begin{pmatrix} x \\ y \\ z \end{pmatrix} \longmapsto \begin{pmatrix} 1 & 0 & 0 \\ 0 & 1 & 0 \end{pmatrix}\begin{pmatrix} x \\ y \\ z \end{pmatrix} = \begin{pmatrix} x \\ y \end{pmatrix}.$

---

2. Example 2.2

$$\begin{array}{ccc}
\boxed{\begin{array}{c} \mathbb{R}^2 \\ \text{standard basis} \\ E = \{(1,0),(0,1)\} \end{array}} & \xrightarrow{\ t\ } & \boxed{\begin{array}{c} \mathbb{R}^4 \\ \text{standard basis} \\ F = \{(1,0,0,0),(0,1,0,0), \\ (0,0,1,0),(0,0,0,1)\} \end{array}}
\end{array}$$

$$t(x,y) = (x,y,x,y)$$

$$t(1,0) = (1,0,1,0), \ t(0,1) = (0,1,0,1).$$

The matrix of t with respect to the standard bases is

$$A = \begin{pmatrix} 1 & 0 \\ 0 & 1 \\ 1 & 0 \\ 0 & 1 \end{pmatrix}.$$

Check: $\begin{pmatrix} x \\ y \end{pmatrix} \longmapsto \begin{pmatrix} 1 & 0 \\ 0 & 1 \\ 1 & 0 \\ 0 & 1 \end{pmatrix}\begin{pmatrix} x \\ y \end{pmatrix} = \begin{pmatrix} x \\ y \\ x \\ y \end{pmatrix}.$

4. Example 2.3

$\mathbb{R}^2$
standard basis
$E = \{(1, 0), (0, 1)\}$

$\xrightarrow{\ \ t\ \ }$

$\mathbb{R}^3$
standard basis
$F = \{(1, 0, 0), (0, 1, 0), (0, 0, 1)\}$

$$t(x, y) = (2x, 3x + y, y)$$

1. $t(1, 0) = (2, 3, 0), \quad t(0, 1) = (0, 1, 1).$
2. $t(1, 0) = (2, 3, 0)_F, \quad t(0, 1) = (0, 1, 1)_F.$
3. $\underline{A} = \begin{pmatrix} 2 & 0 \\ 3 & 1 \\ 0 & 1 \end{pmatrix}.$

Thus $\begin{pmatrix} x \\ y \end{pmatrix} \longmapsto \begin{pmatrix} 2 & 0 \\ 3 & 1 \\ 0 & 1 \end{pmatrix} \begin{pmatrix} x \\ y \end{pmatrix} = \begin{pmatrix} 2x \\ 3x + y \\ y \end{pmatrix}.$

we often omit the subscripts for standard bases

6. Exercise 2.4

For each of the following linear transformations t, find the matrix representation of t with respect to the standard bases for the domain and codomain.

(a) $t: \mathbb{R}^2 \longrightarrow \mathbb{R}^2$
$(x, y) \longmapsto (x + 3y, y)$

(b) $t: P_3 \longrightarrow P_3$
$p(x) \longmapsto p(x) + p(2)$

use the strategy

5. Example 2.4

P_3
standard basis
$E = \{1, x, x^2\}$

$\xrightarrow{\ \ t\ \ }$

P_2
standard basis
$F = \{1, x\}$

$$t(p(x)) = p'(x)$$
$$t(a + bx + cx^2) = b + 2cx.$$

1. $t(1) = 0, \quad t(x) = 1, \quad t(x^2) = 2x.$
2. $t(1) = 0 \cdot 1 + 0 \cdot x = (0, 0)_F, \quad t(x) = 1 \cdot 1 + 0 \cdot x = (1, 0)_F,$
$t(x^2) = 0 \cdot 1 + 2 \cdot x = (0, 2)_F.$
3. $\underline{A} = \begin{pmatrix} 0 & 1 & 0 \\ 0 & 0 & 2 \end{pmatrix}.$

Thus $\begin{pmatrix} a \\ b \\ c \end{pmatrix}_E \longmapsto \begin{pmatrix} 0 & 1 & 0 \\ 0 & 0 & 2 \end{pmatrix} \begin{pmatrix} a \\ b \\ c \end{pmatrix}_E = \begin{pmatrix} b \\ 2c \end{pmatrix}_F.$

7. Non-standard domain basis

$\mathbb{R}^3$
non-standard basis
$E = \{(1, 1, 1), (1, 1, 0), (1, 0, 0)\}$

$\xrightarrow{\ \ t\ \ }$

$\mathbb{R}^2$
standard basis
$F = \{(1, 0), (0, 1)\}$

$$t(x, y, z) = (x, y)$$

1. $t(1, 1, 1) = (1, 1), \quad t(1, 1, 0) = (1, 1), \quad t(1, 0, 0) = (1, 0).$
2. $t(1, 1, 1) = (1, 1)_F, \quad t(1, 1, 0) = (1, 1)_F, \quad t(1, 0, 0) = (1, 0)_F.$
3. $\underline{A} = \begin{pmatrix} 1 & 1 & 1 \\ 1 & 1 & 0 \end{pmatrix}.$

Thus $\begin{pmatrix} v_1 \\ v_2 \\ v_3 \end{pmatrix}_E \longmapsto \begin{pmatrix} 1 & 1 & 1 \\ 1 & 1 & 0 \end{pmatrix} \begin{pmatrix} v_1 \\ v_2 \\ v_3 \end{pmatrix}_E = \begin{pmatrix} v_1 + v_2 + v_3 \\ v_1 + v_2 \end{pmatrix}_F.$

Note that different bases give different matrix representations.

compare Frame 1

9. Non-standard bases

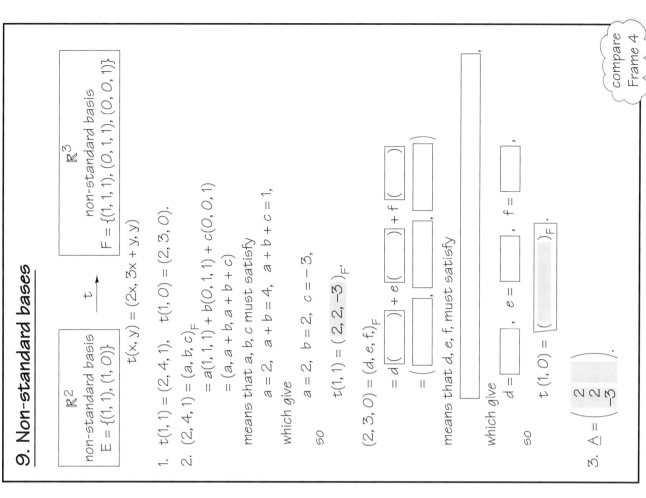

$\mathbb{R}^2$ non-standard basis $E = \{(1,1), (1,0)\}$

$\mathbb{R}^3$ non-standard basis $F = \{(1,1,1), (0,1,1), (0,0,1)\}$

$$t(x,y) = (2x, 3x+y, y)$$

1. $t(1,1) = (2,4,1)$, $t(1,0) = (2,3,0)$.

2. $(2,4,1) = (a,b,c)_F$
   $= a(1,1,1) + b(0,1,1) + c(0,0,1)$
   $= (a, a+b, a+b+c)$

   means that a, b, c must satisfy
   $a=2$, $a+b=4$, $a+b+c=1$,

   which give
   $a=2$, $b=2$, $c=-3$,

   so $t(1,1) = (2, 2, -3)_F$.

   $(2,3,0) = (d,e,f)_F$
   $= d(\quad) + e(\quad) + f(\quad)$
   $= (\quad, \quad, \quad)$

   means that d, e, f must satisfy

   which give
   $d = \quad$, $e = \quad$, $f = \quad$,

   so $t(1,0) = (\quad)_F$.

3. $\underline{A} = \begin{pmatrix} 2 \\ 2 \\ -3 \end{pmatrix}$.

compare Frame 4

8. Non-standard codomain basis

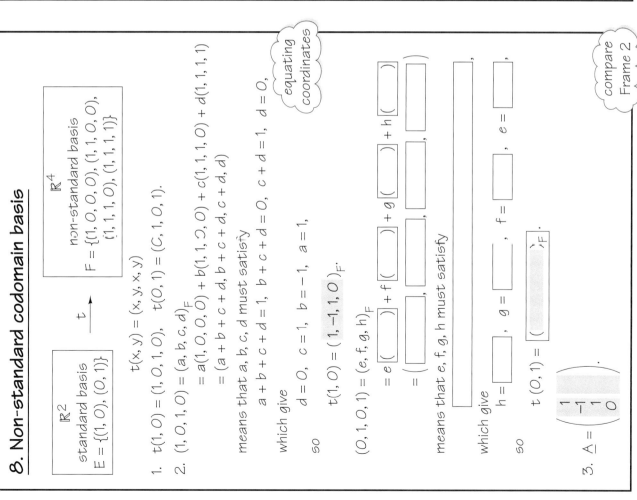

$\mathbb{R}^2$ standard basis $E = \{(1,0), (0,1)\}$

$\mathbb{R}^4$ non-standard basis $F = \{(1,0,0,0), (1,1,0,0), (1,1,1,0), (1,1,1,1)\}$

$$t(x,y) = (x, y, x, y)$$

1. $t(1,0) = (1,0,1,0)$, $t(0,1) = (0,1,0,1)$.

2. $(1,0,1,0) = (a,b,c,d)_F$
   $= a(1,0,0,0) + b(1,1,0,0) + c(1,1,1,0) + d(1,1,1,1)$
   $= (a+b+c+d, b+c+d, c+d, d)$

   means that a, b, c, d must satisfy
   $a+b+c+d=1$, $b+c+d=0$, $c+d=1$, $d=0$,   *equating coordinates*

   which give
   $d=0$, $c=1$, $b=-1$, $a=1$,

   so $t(1,0) = (1, -1, 1, 0)_F$.

   $(0,1,0,1) = (e,f,g,h)_F$
   $= e(\quad) + f(\quad) + g(\quad) + h(\quad)$
   $= (\quad, \quad, \quad)$

   means that e, f, g, h must satisfy

   which give
   $h = \quad$, $g = \quad$, $f = \quad$, $e = \quad$,

   so $t(0,1) = (\quad)_F$.

3. $\underline{A} = \begin{pmatrix} 1 \\ -1 \\ 1 \\ 0 \end{pmatrix}$.

compare Frame 2

Post-audio exercises

Exercise 2.5   Find the matrix of the linear transformation

$$t : \mathbb{R}^3 \longrightarrow \mathbb{R}^2$$

$$(x, y, z) \longmapsto (x, y)$$

with respect to each of the following bases E for $\mathbb{R}^3$ and F for $\mathbb{R}^2$.

(a) $E = \{(1, 0, 1), (1, 0, 0), (1, 1, 1)\}$
     $F = \{(1, 0), (0, 1)\}$ (standard basis for $\mathbb{R}^2$)

(b) $E = \{(1, 0, 0), (0, 1, 0), (0, 0, 1)\}$ (standard basis for $\mathbb{R}^3$)
     $F = \{(2, 1), (1, 1)\}$

(c) $E = \{(0, 1, 0), (1, 1, 1), (0, 1, 1)\}$ $F = \{(1, 3), (2, 4)\}$

In Exercise 2.5 you saw that a linear transformation $t : V \longrightarrow W$ has many different matrix representations:

- *different bases* for V and W give *different* matrix representations.

Moreover, the order of the elements in a basis is important:

- a *different order* gives a *different* matrix representation.

For example, you should obtain different answers to Exercise 2.6(a) and (b): although the bases contain the same elements, the order in which they appear in the domain basis is different.

Exercise 2.6   Find the matrix representation of the linear transformation

$$t : P_3 \longrightarrow P_2$$

$$p(x) \longmapsto p'(x)$$

with respect to each of the following bases E for P_3 and F for P_2.

(a) $E = \{1, x, x^2\}$ $F = \{2x, 1 + x\}$

(b) $E = \{x, x^2, 1\}$ $F = \{2x, 1 + x\}$

We now prove that for a given linear transformation $t : V \longrightarrow W$ and given (ordered) bases for V and W, there is precisely *one* matrix representation—the one given by Strategy 2.1 in Frame 3. The proof uses the fact that linear transformations preserve linear combinations of vectors.

Theorem 2.1   Let $t : V \longrightarrow W$ be a linear transformation, let $E = \{\mathbf{e}_1, \ldots, \mathbf{e}_n\}$ be a basis for V and let $F = \{\mathbf{f}_1, \ldots, \mathbf{f}_m\}$ be a basis for W. Let

$$t(\mathbf{e}_1) = (a_{11}, a_{21}, \ldots, a_{m1})_F,$$
$$t(\mathbf{e}_2) = (a_{12}, a_{22}, \ldots, a_{m2})_F,$$
$$\vdots$$
$$t(\mathbf{e}_n) = (a_{1n}, a_{2n}, \ldots, a_{mn})_F.$$

Then there is exactly one matrix of t with respect to the bases E and F, namely

$$\mathbf{A} = \begin{pmatrix} a_{11} & a_{12} & \cdots & a_{1n} \\ a_{21} & a_{22} & \cdots & a_{2n} \\ \vdots & \vdots & & \vdots \\ a_{m1} & a_{m2} & \cdots & a_{mn} \end{pmatrix}.$$

Remark   It follows that the unique matrix representation of t with respect to the bases E and F is

$$\begin{pmatrix} v_1 \\ v_2 \\ \vdots \\ v_n \end{pmatrix}_E \longmapsto \begin{pmatrix} a_{11} & a_{12} & \cdots & a_{1n} \\ a_{21} & a_{22} & \cdots & a_{2n} \\ \vdots & \vdots & & \vdots \\ a_{m1} & a_{m2} & \cdots & a_{mn} \end{pmatrix} \begin{pmatrix} v_1 \\ v_2 \\ \vdots \\ v_n \end{pmatrix}_E$$

$$= \begin{pmatrix} a_{11}v_1 + \cdots + a_{1n}v_n \\ a_{21}v_1 + \cdots + a_{2n}v_n \\ \vdots \\ a_{m1}v_1 + \cdots + a_{mn}v_n \end{pmatrix}_F .$$

Proof   Suppose that the conditions of the theorem are satisfied and that $(v_1, \ldots, v_n)_E$ is the E-coordinate representation of a vector $\mathbf{v} \in V$. Then we have

$$\mathbf{v} = v_1\mathbf{e}_1 + v_2\mathbf{e}_2 + \cdots + v_n\mathbf{e}_n.$$

If you are short of time, omit this proof.

Linear transformations preserve linear combinations of vectors, so

See Theorem 1.3, page 16.

$$\begin{aligned} t(\mathbf{v}) &= v_1 t(\mathbf{e}_1) + v_2 t(\mathbf{e}_2) + \cdots + v_n t(\mathbf{e}_n) \\ &= v_1(a_{11}, a_{21}, \ldots, a_{m1})_F + v_2(a_{12}, a_{22}, \ldots, a_{m2})_F + \cdots \\ &\quad + v_n(a_{1n}, a_{2n}, \ldots, a_{mn})_F \\ &= (v_1 a_{11} + \cdots + v_n a_{1n}, \; v_1 a_{21} + \cdots + v_n a_{2n}, \; \ldots, \\ &\quad v_1 a_{m1} + \cdots + v_n a_{mn})_F. \end{aligned}$$

So the first coordinate of $t(\mathbf{v})$ is $a_{11}v_1 + \cdots + a_{1n}v_n$; this is the dot product

$$(a_{11}, \ldots, a_{1n}) \cdot (v_1, \ldots, v_n).$$

The second coordinate of $t(\mathbf{v})$ is $a_{21}v_1 + \cdots + a_{2n}v_n$; this is the dot product

$$(a_{21}, \ldots, a_{2n}) \cdot (v_1, \ldots, v_n).$$

Continuing in this way, we find that

$$\begin{pmatrix} a_{11}v_1 + \cdots + a_{1n}v_n \\ a_{21}v_1 + \cdots + a_{2n}v_n \\ \vdots \\ a_{m1}v_1 + \cdots + a_{mn}v_n \end{pmatrix} = \begin{pmatrix} a_{11} & a_{12} & \cdots & a_{1n} \\ a_{21} & a_{22} & \cdots & a_{2n} \\ \vdots & \vdots & & \vdots \\ a_{m1} & a_{m2} & \cdots & a_{mn} \end{pmatrix} \begin{pmatrix} v_1 \\ v_2 \\ \vdots \\ v_n \end{pmatrix} .$$

It follows that

$$\begin{pmatrix} v_1 \\ v_2 \\ \vdots \\ v_n \end{pmatrix}_E \longmapsto \begin{pmatrix} a_{11} & a_{12} & \cdots & a_{1n} \\ a_{21} & a_{22} & \cdots & a_{2n} \\ \vdots & \vdots & & \vdots \\ a_{m1} & a_{m2} & \cdots & a_{mn} \end{pmatrix} \begin{pmatrix} v_1 \\ v_2 \\ \vdots \\ v_n \end{pmatrix}_E$$

$$= \begin{pmatrix} a_{11}v_1 + \cdots + a_{1n}v_n \\ a_{21}v_1 + \cdots + a_{2n}v_n \\ \vdots \\ a_{m1}v_1 + \cdots + a_{mn}v_n \end{pmatrix}_F$$

is a matrix representation of t with respect to the bases E and F; that is,

$$\mathbf{A} = \begin{pmatrix} a_{11} & a_{12} & \cdots & a_{1n} \\ a_{21} & a_{22} & \cdots & a_{2n} \\ \vdots & \vdots & & \vdots \\ a_{m1} & a_{m2} & \cdots & a_{mn} \end{pmatrix}$$

is a matrix of t with respect to the bases E and F.

It remains to show that $\mathbf{A}$ is the *only* possible matrix of t with respect to the bases E and F. Suppose that

$$\mathbf{B} = \begin{pmatrix} b_{11} & b_{12} & \cdots & b_{1n} \\ b_{21} & b_{22} & \cdots & b_{2n} \\ \vdots & \vdots & & \vdots \\ b_{m1} & b_{m2} & \cdots & b_{mn} \end{pmatrix}$$

is also a matrix of t with respect to the bases E and F. Then

$$\begin{pmatrix} 1 \\ 0 \\ \vdots \\ 0 \end{pmatrix}_E \longmapsto \begin{pmatrix} b_{11} & b_{12} & \cdots & b_{1n} \\ b_{21} & b_{22} & \cdots & b_{2n} \\ \vdots & \vdots & & \vdots \\ b_{m1} & b_{m2} & \cdots & b_{mn} \end{pmatrix} \begin{pmatrix} 1 \\ 0 \\ \vdots \\ 0 \end{pmatrix}_E = \begin{pmatrix} b_{11} \\ b_{21} \\ \vdots \\ b_{m1} \end{pmatrix}_F,$$

so, as $\mathbf{e}_1 = (1, 0, \ldots, 0)_E$, we have

$$t(\mathbf{e}_1) = (b_{11}, b_{21}, \ldots, b_{m1})_F.$$

Similarly, we find that

$$t(\mathbf{e}_2) = (b_{12}, b_{22}, \ldots, b_{m2})_F,$$
$$\vdots$$
$$t(\mathbf{e}_n) = (b_{1n}, b_{2n}, \ldots, b_{mn})_F.$$

But

$$t(\mathbf{e}_1) = (a_{11}, a_{21}, \ldots, a_{m1})_F,$$
$$t(\mathbf{e}_2) = (a_{12}, a_{22}, \ldots, a_{m2})_F,$$
$$\vdots$$
$$t(\mathbf{e}_n) = (a_{1n}, a_{2n}, \ldots, a_{mn})_F,$$

so each column of $\mathbf{B}$ is the same as the corresponding column of $\mathbf{A}$. Since $\mathbf{A}$ and $\mathbf{B}$ have the same number of columns, it follows that $\mathbf{B} = \mathbf{A}$. Thus $\mathbf{A}$ is indeed the only matrix of t with respect to the bases E and F. ∎

2.2 An equivalent definition of a linear transformation

We have shown that any linear transformation $t : V \longrightarrow W$, where V and W are finite-dimensional vector spaces, has a matrix representation. We now show the converse, that a function which has a matrix representation is a linear transformation.

Theorem 2.2   Let $t : V \longrightarrow W$ be a function which has a matrix representation. Then t is a linear transformation.

Proof   Suppose that the function $t : V \longrightarrow W$ has a matrix representation

$$\mathbf{v}_E \longmapsto \mathbf{A}\mathbf{v}_E = t(\mathbf{v})_F.$$

We first show that t satisfies LT1:

$$t(\mathbf{v}_1 + \mathbf{v}_2) = t(\mathbf{v}_1) + t(\mathbf{v}_2), \quad \text{for all } \mathbf{v}_1, \mathbf{v}_2 \in V.$$

If you are short of time, omit this proof.

Let $\mathbf{v}_1, \mathbf{v}_2 \in V$. Then

$$t(\mathbf{v}_1 + \mathbf{v}_2)_F = \mathbf{A}(\mathbf{v}_1 + \mathbf{v}_2)_E$$

and

$$t(\mathbf{v}_1)_F + t(\mathbf{v}_2)_F = \mathbf{A}(\mathbf{v}_1)_E + \mathbf{A}(\mathbf{v}_2)_E = \mathbf{A}(\mathbf{v}_1 + \mathbf{v}_2)_E.$$

So $t(\mathbf{v}_1 + \mathbf{v}_2)_F = t(\mathbf{v}_1)_F + t(\mathbf{v}_2)_F$, and hence $t(\mathbf{v}_1 + \mathbf{v}_2) = t(\mathbf{v}_1) + t(\mathbf{v}_2)$. Thus LT1 is satisfied.

We now show that t satisfies LT2:

$$t(\alpha\mathbf{v}) = \alpha\, t(\mathbf{v}), \quad \text{for all } \mathbf{v} \in V, \ \alpha \in \mathbb{R}.$$

Let $\mathbf{v} \in V$ and $\alpha \in \mathbb{R}$. Then

$$\alpha\, t(\mathbf{v})_F = \alpha \mathbf{A} \mathbf{v}_E$$

and

$$t(\alpha\mathbf{v})_F = \mathbf{A}(\alpha\mathbf{v})_E = \alpha \mathbf{A} \mathbf{v}_E.$$

So $t(\alpha\mathbf{v})_F = \alpha\, t(\mathbf{v})_F$, and hence $t(\alpha\mathbf{v}) = \alpha\, t(\mathbf{v})$. Thus LT2 is satisfied.

Since LT1 and LT2 are satisfied, t is a linear transformation. ■

Theorems 2.1 and 2.2 show that

- the linear transformations from a finite-dimensional vector space V to a finite-dimensional vector space W are precisely those functions from V to W that have a matrix representation.

This means, for example, that the linear transformations from $\mathbb{R}^2$ to itself are those functions that have a matrix representation

$$\begin{pmatrix} x \\ y \end{pmatrix} \longmapsto \begin{pmatrix} a & b \\ c & d \end{pmatrix} \begin{pmatrix} x \\ y \end{pmatrix} = \begin{pmatrix} ax + by \\ cx + dy \end{pmatrix}.$$

So the linear transformations from $\mathbb{R}^2$ to itself are those functions of the form

$$t : \mathbb{R}^2 \longrightarrow \mathbb{R}^2$$
$$(x, y) \longmapsto (ax + by, cx + dy)$$

for some $a, b, c, d \in \mathbb{R}$.

Similar expressions exist for linear transformations from $\mathbb{R}^n$ to $\mathbb{R}^m$.

Exercise 2.7   State which of the following functions are linear transformations.

(a) $t : \mathbb{R}^2 \longrightarrow \mathbb{R}^2$
$(x, y) \longmapsto (y, 2x + y)$

(b) $t : \mathbb{R}^2 \longrightarrow \mathbb{R}^2$
$(x, y) \longmapsto (x^2, y)$

(c) $t : \mathbb{R}^2 \longrightarrow \mathbb{R}^2$
$(x, y) \longmapsto (x, 2xy + y)$

(d) $t : \mathbb{R}^2 \longrightarrow \mathbb{R}^2$
$(x, y) \longmapsto (3x, x + 4y)$

Further exercises

Exercise 2.8   For each of the following linear transformations, find the matrix of t with respect to the standard bases for the domain and codomain.

(a) $t : \mathbb{R}^2 \longrightarrow \mathbb{R}^2$
$(x, y) \longmapsto (3x + y, 2x - y)$

(b) $t : \mathbb{R}^3 \longrightarrow \mathbb{R}^2$
$(x, y, z) \longmapsto (x + y, x - z)$

Exercise 2.9   Find the matrix representations of the linear transformation

$$t : \mathbb{R}^2 \longrightarrow \mathbb{R}^2$$
$$(x, y) \longmapsto (x + y, 2x + 3y)$$

with respect to the following bases E for the domain and F for the codomain.

(a) $E = \{(1, 1), (0, 1)\}$ $F = \{(1, 0), (0, 1)\}$

(b) $E = \{(1, 0), (0, 1)\}$ $F = \{(2, 1), (1, 3)\}$

(c) $E = \{(1, 2), (1, 1)\}$ $F = \{(1, 0), (3, 1)\}$

Exercise 2.10   Let $V = \{f(x) : f(x) = ae^x \cos x + be^x \sin x, \ a, b \in \mathbb{R}\}$. Find the matrix of the linear transformation

$$t : V \longrightarrow V$$
$$f(x) \longmapsto f'(x)$$

with respect to the basis $\{e^x \cos x, e^x \sin x\}$ for both the domain and codomain.

Exercise 2.11   Find the matrix representations of the linear transformation

$$t : P_3 \longrightarrow P_3$$
$$p(x) \longmapsto p(x) + p'(x)$$

with respect to the following bases E for the domain and F for the codomain.

(a) $E = F = \{1, x, x^2\}$

(b) $E = \{1 + x, x, x^2\}$ $F = \{1, x, x^2\}$

(c) $E = \{1, x, x^2\}$ $F = \{1 + x, x, x^2\}$

3   Composition and invertibility

After working through this section, you should be able to:

(a) use the matrix representations of two given linear transformations s and t to find a matrix representation of the composite function $s \circ t$;

(b) determine whether a given linear transformation is invertible and, if it is, find its inverse;

(c) understand that each n-dimensional vector space is isomorphic to $\mathbb{R}^n$.

3.1   Composition Rule

In the previous section you saw that a function $t : V \longrightarrow W$, where V and W are finite-dimensional vector spaces, is a linear transformation if and only if it has a matrix representation. We now use some of the properties of matrices that you met in Unit LA2 to develop our understanding of linear transformations.

We begin by considering the composition of linear transformations. The composite of two functions $t : V \longrightarrow W$ and $s : W \longrightarrow X$ is

Composite functions were introduced in Unit I2.

$$s \circ t : V \longrightarrow X$$
$$\mathbf{v} \longmapsto s(t(\mathbf{v})).$$

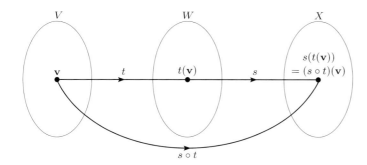

Consider the linear transformations

$$t : \mathbb{R}^2 \longrightarrow \mathbb{R}^2 \qquad s : \mathbb{R}^2 \longrightarrow \mathbb{R}^2$$
$$(x, y) \longmapsto (x + 2y, y) \quad \text{and} \quad (x, y) \longmapsto (5x, x + y). \tag{3.1}$$

Let $(x, y) \in \mathbb{R}^2$. Then

$$t(x, y) = (x + 2y, y),$$

so

$$s(t(x, y)) = s(x + 2y, y)$$
$$= (5(x + 2y), (x + 2y) + y)$$
$$= (5x + 10y, x + 3y).$$

Thus the composite function $s \circ t$ is the linear transformation

$$s \circ t : \mathbb{R}^2 \longrightarrow \mathbb{R}^2$$
$$(x, y) \longmapsto (5x + 10y, x + 3y).$$

In general, for linear transformations s and t from a vector space to itself, the composite functions $s \circ t$ and $t \circ s$ are not the same, as you will see in the following exercise.

Exercise 3.1   Let t and s be the linear transformations

$$t : \mathbb{R}^2 \longrightarrow \mathbb{R}^2 \qquad s : \mathbb{R}^2 \longrightarrow \mathbb{R}^2$$
$$(x, y) \longmapsto (3x + y, -x) \quad \text{and} \quad (x, y) \longmapsto (x, x + y).$$

Find the following composite functions.

(a) $s \circ t$ (b) $t \circ s$

Each of the composite functions in Exercise 3.1 is a linear transformation. In our next theorem (Theorem 3.1) we show that composition of two linear transformations gives a linear transformation.

At the beginning of this section we showed that the two linear transformations s and t in equation (3.1) can be composed to give the linear transformation

$$s \circ t : \mathbb{R}^2 \longrightarrow \mathbb{R}^2$$
$$(x, y) \longmapsto (5x + 10y, x + 3y).$$

Using Strategy 2.1 we obtain the matrix representations of these three linear transformations with respect to the standard basis for $\mathbb{R}^2$:

$$t : \mathbb{R}^2 \longrightarrow \mathbb{R}^2$$
$$\begin{pmatrix} x \\ y \end{pmatrix} \longmapsto \begin{pmatrix} 1 & 2 \\ 0 & 1 \end{pmatrix} \begin{pmatrix} x \\ y \end{pmatrix} = \begin{pmatrix} x + 2y \\ y \end{pmatrix},$$

$$s : \mathbb{R}^2 \longrightarrow \mathbb{R}^2$$
$$\begin{pmatrix} x \\ y \end{pmatrix} \longmapsto \begin{pmatrix} 5 & 0 \\ 1 & 1 \end{pmatrix} \begin{pmatrix} x \\ y \end{pmatrix} = \begin{pmatrix} 5x \\ x + y \end{pmatrix}$$

and

$$s \circ t : \mathbb{R}^2 \longrightarrow \mathbb{R}^2$$
$$\begin{pmatrix} x \\ y \end{pmatrix} \longmapsto \begin{pmatrix} 5 & 10 \\ 1 & 3 \end{pmatrix} \begin{pmatrix} x \\ y \end{pmatrix} = \begin{pmatrix} 5x + 10y \\ x + 3y \end{pmatrix}.$$

We can check that

$$\begin{pmatrix} 5 & 10 \\ 1 & 3 \end{pmatrix} = \begin{pmatrix} 5 & 0 \\ 1 & 1 \end{pmatrix} \begin{pmatrix} 1 & 2 \\ 0 & 1 \end{pmatrix},$$

so, in this example,

$$\begin{pmatrix} \text{matrix} \\ \text{of } s \circ t \end{pmatrix} = \begin{pmatrix} \text{matrix} \\ \text{of } s \end{pmatrix} \begin{pmatrix} \text{matrix} \\ \text{of } t \end{pmatrix}.$$

We now show that this relationship between the matrices of $s \circ t$, s and t holds in general.

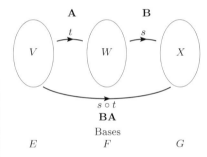

Theorem 3.1 Composition Rule

Let $t : V \longrightarrow W$ and $s : W \longrightarrow X$ be linear transformations. Then:

(a) $s \circ t : V \longrightarrow X$ is a linear transformation;

(b) if $\mathbf{A}$ is the matrix of t with respect to the bases E and F, and $\mathbf{B}$ is the matrix of s with respect to the bases F and G, then $\mathbf{BA}$ is the matrix of $s \circ t$ with respect to the bases E and G.

Remark   The apparently 'strange' rule for multiplying matrices was designed so that the composition of linear transformations corresponds to the multiplication of matrices.

Proof   Let $t : V \longrightarrow W$ and $s : W \longrightarrow X$ be linear transformations.

(a) First we show that $s \circ t$ satisfies LT1:

$$(s \circ t)(\mathbf{v}_1 + \mathbf{v}_2) = (s \circ t)(\mathbf{v}_1) + (s \circ t)(\mathbf{v}_2), \quad \text{for all } \mathbf{v}_1, \mathbf{v}_2 \in V.$$

Let $\mathbf{v}_1, \mathbf{v}_2 \in V$. Then, since t and s both satisfy LT1, we have

$$\begin{aligned} (s \circ t)(\mathbf{v}_1 + \mathbf{v}_2) &= s(t(\mathbf{v}_1 + \mathbf{v}_2)) \\ &= s(t(\mathbf{v}_1) + t(\mathbf{v}_2)) \\ &= s(t(\mathbf{v}_1)) + s(t(\mathbf{v}_2)). \end{aligned}$$

We also have

$$(s \circ t)(\mathbf{v}_1) + (s \circ t)(\mathbf{v}_2) = s(t(\mathbf{v}_1)) + s(t(\mathbf{v}_2)).$$

These expressions are equal, so LT1 is satisfied.

Next we show that $s \circ t$ satisfies LT2:

$$(s \circ t)(\alpha \mathbf{v}) = \alpha(s \circ t)(\mathbf{v}), \quad \text{for all } \mathbf{v} \in V, \ \alpha \in \mathbb{R}.$$

If you are short of time, omit this proof.

Let $\mathbf{v} \in V$ and $\alpha \in \mathbb{R}$. Then, since t and s both satisfy LT2, we have

$$(s \circ t)(\alpha \mathbf{v}) = s(t(\alpha \mathbf{v})) = s(\alpha\, t(\mathbf{v})) = \alpha\, s(t(\mathbf{v})).$$

We also have

$$\alpha(s \circ t)(\mathbf{v}) = \alpha\, s(t(\mathbf{v})).$$

These expressions are equal, so LT2 is satisfied.

Since LT1 and LT2 are satisfied, $s \circ t$ is a linear transformation.

(b) Suppose that the vector spaces V, W and X have dimensions n, m and p, respectively. Then $\mathbf{A}$ is an $m \times n$ matrix of the form

$$\mathbf{A} = \begin{pmatrix} a_{11} & a_{12} & \cdots & a_{1n} \\ a_{21} & a_{22} & \cdots & a_{2n} \\ \vdots & \vdots & & \vdots \\ a_{m1} & a_{m2} & \cdots & a_{mn} \end{pmatrix}$$

and $\mathbf{B}$ is a $p \times m$ matrix of the form

$$\mathbf{B} = \begin{pmatrix} b_{11} & b_{12} & \cdots & b_{1m} \\ b_{21} & b_{22} & \cdots & b_{2m} \\ \vdots & \vdots & & \vdots \\ b_{p1} & b_{p2} & \cdots & b_{pm} \end{pmatrix}.$$

We use Strategy 2.1 to find the matrix of the linear transformation $s \circ t$ with respect to the bases E and G.

First we find the images under $s \circ t$ of the vectors $\mathbf{e}_1, \dots, \mathbf{e}_n$ which form the basis E for V. To find the image of the basis vector $\mathbf{e}_1$, we use the $n \times 1$ column matrix containing the coordinates of $\mathbf{e}_1$ with respect to the basis E. This matrix has 1 in the first row and 0 elsewhere. Using the matrix representations of t and s, we find that

$$t: \begin{pmatrix} 1 \\ 0 \\ \vdots \\ 0 \end{pmatrix}_E \longmapsto \begin{pmatrix} a_{11} & a_{12} & \cdots & a_{1n} \\ a_{21} & a_{22} & \cdots & a_{2n} \\ \vdots & \vdots & & \vdots \\ a_{m1} & a_{m2} & \cdots & a_{mn} \end{pmatrix} \begin{pmatrix} 1 \\ 0 \\ \vdots \\ 0 \end{pmatrix}_E = \begin{pmatrix} a_{11} \\ a_{21} \\ \vdots \\ a_{m1} \end{pmatrix}_F$$

and

$$s: \begin{pmatrix} a_{11} \\ a_{21} \\ \vdots \\ a_{m1} \end{pmatrix}_F \longmapsto \begin{pmatrix} b_{11} & b_{12} & \cdots & b_{1m} \\ b_{21} & b_{22} & \cdots & b_{2m} \\ \vdots & \vdots & & \vdots \\ b_{p1} & b_{p2} & \cdots & b_{pm} \end{pmatrix} \begin{pmatrix} a_{11} \\ a_{21} \\ \vdots \\ a_{m1} \end{pmatrix}_F$$

$$= \begin{pmatrix} b_{11}a_{11} + \cdots + b_{1m}a_{m1} \\ b_{21}a_{11} + \cdots + b_{2m}a_{m1} \\ \vdots \\ b_{p1}a_{11} + \cdots + b_{pm}a_{m1} \end{pmatrix}_G.$$

So

$$(s \circ t)(\mathbf{e}_1) = (b_{11}a_{11} + \cdots + b_{1m}a_{m1}, \dots, b_{p1}a_{11} + \cdots + b_{pm}a_{m1})_G.$$

Similarly, we find that, for $k = 2, \dots, n$,

$$(s \circ t)(\mathbf{e}_k) = (b_{11}a_{1k} + \cdots + b_{1m}a_{mk}, \dots, b_{p1}a_{1k} + \cdots + b_{pm}a_{mk})_G.$$

The next step of Strategy 2.1 is to find the G-coordinates of each of the image vectors, but the image vectors are already in this form, so we go on to the third step.

The third step of Strategy 2.1 is to construct the matrix of $s \circ t$, column by column. The first column contains the coordinates of $(s \circ t)(\mathbf{e}_1)$, the second column contains the coordinates of $(s \circ t)(\mathbf{e}_2)$, and so on. Thus the matrix of $s \circ t$ with respect to the bases E and G is

$$\begin{pmatrix} b_{11}a_{11} + \cdots + b_{1m}a_{m1} & \cdots & b_{11}a_{1k} + \cdots + b_{1m}a_{mk} & \cdots & b_{11}a_{1n} + \cdots + b_{1m}a_{mn} \\ \vdots & & \vdots & & \vdots \\ b_{j1}a_{11} + \cdots + b_{jm}a_{m1} & \cdots & b_{j1}a_{1k} + \cdots + b_{jm}a_{mk} & \cdots & b_{j1}a_{1n} + \cdots + b_{jm}a_{mn} \\ \vdots & & \vdots & & \vdots \\ b_{p1}a_{11} + \cdots + b_{pm}a_{m1} & \cdots & b_{p1}a_{1k} + \cdots + b_{pm}a_{mk} & \cdots & b_{p1}a_{1n} + \cdots + b_{pm}a_{mn} \end{pmatrix}.$$

Using the rules for matrix multiplication, we find that the above matrix is the same as the matrix product $\mathbf{BA}$. Thus $\mathbf{BA}$ is the matrix of $s \circ t$ with respect to the bases E and G.   ■

Exercise 3.2   In each of the following cases, use the Composition Rule to find the matrix representation of the linear transformation $s \circ t$ with respect to the standard bases for the domain and codomain.

(a) $t : \mathbb{R}^3 \longrightarrow \mathbb{R}^2$

$$\begin{pmatrix} x \\ y \\ z \end{pmatrix} \longmapsto \begin{pmatrix} 2 & 1 & 0 \\ 0 & 1 & 3 \end{pmatrix} \begin{pmatrix} x \\ y \\ z \end{pmatrix}$$

$s : \mathbb{R}^2 \longrightarrow \mathbb{R}^2$

$$\begin{pmatrix} x \\ y \end{pmatrix} \longmapsto \begin{pmatrix} 1 & 2 \\ 4 & 3 \end{pmatrix} \begin{pmatrix} x \\ y \end{pmatrix}$$

(b) $t : \mathbb{R}^4 \longrightarrow \mathbb{R}^2$

$$\begin{pmatrix} x \\ y \\ z \\ w \end{pmatrix} \longmapsto \begin{pmatrix} 1 & 0 & 2 & 4 \\ 2 & 1 & 0 & 3 \end{pmatrix} \begin{pmatrix} x \\ y \\ z \\ w \end{pmatrix}$$

$s : \mathbb{R}^2 \longrightarrow \mathbb{R}^3$

$$\begin{pmatrix} x \\ y \end{pmatrix} \longmapsto \begin{pmatrix} 2 & 1 \\ 0 & 2 \\ 1 & 0 \end{pmatrix} \begin{pmatrix} x \\ y \end{pmatrix}$$

We now return to two examples of linear transformations of vector spaces of polynomials:

$$t : P_3 \longrightarrow P_3$$
$$p(x) \longmapsto p(x) + p(2)$$

and

$$s : P_3 \longrightarrow P_2$$
$$p(x) \longmapsto p'(x).$$

We compose these linear transformations as follows:

$$\begin{aligned} (s \circ t)(p(x)) &= s(t(p(x))) \\ &= s(p(x) + p(2)) \\ &= (p(x) + p(2))' \\ &= p'(x). \end{aligned}$$

In this case, the functions $s \circ t$ and s are the same function.

Thus the composite is

$$s \circ t : P_3 \longrightarrow P_2$$
$$p(x) \longmapsto p'(x).$$

Exercise 3.3   Use the Composition Rule to find a matrix representation of the linear transformation $s \circ t$ when

$$t : P_3 \longrightarrow P_3 \qquad s : P_3 \longrightarrow P_2$$
$$p(x) \longmapsto p(x) + p(2) \quad \text{and} \quad p(x) \longmapsto p'(x).$$

(Recall that s and t have the matrix representations

See the solution to
Exercise 2.4(b) for t and
Example 2.4 in Frame 5 for s.

$$t : P_3 \longrightarrow P_3$$

$$\begin{pmatrix} a \\ b \\ c \end{pmatrix}_E \longmapsto \begin{pmatrix} 2 & 2 & 4 \\ 0 & 1 & 0 \\ 0 & 0 & 1 \end{pmatrix} \begin{pmatrix} a \\ b \\ c \end{pmatrix}_E = \begin{pmatrix} 2a + 2b + 4c \\ b \\ c \end{pmatrix}_F$$

and

$$s : P_3 \longrightarrow P_2$$

$$\begin{pmatrix} a \\ b \\ c \end{pmatrix}_E \longmapsto \begin{pmatrix} 0 & 1 & 0 \\ 0 & 0 & 2 \end{pmatrix} \begin{pmatrix} a \\ b \\ c \end{pmatrix}_E = \begin{pmatrix} b \\ 2c \end{pmatrix}_F,$$

where $E = \{1, x, x^2\}$ and $F = \{1, x\}$ are the standard bases for P_3 and P_2, respectively.)

We claimed earlier that multiplication of matrices is associative. We now prove this result, by using the Composition Rule.

See Unit LA2, Section 3.

Corollary to Theorem 3.1

Let $\mathbf{A}$, $\mathbf{B}$ and $\mathbf{C}$ be matrices of sizes $q \times p$, $p \times m$ and $m \times n$, respectively. Then

$$\mathbf{A}(\mathbf{B}\mathbf{C}) = (\mathbf{A}\mathbf{B})\mathbf{C}.$$

Proof   Let t, s and r be the linear transformations whose matrix representations with respect to the standard bases for the domain and codomain are

$$t : \mathbb{R}^n \longrightarrow \mathbb{R}^m \qquad s : \mathbb{R}^m \longrightarrow \mathbb{R}^p \qquad \text{and} \qquad r : \mathbb{R}^p \longrightarrow \mathbb{R}^q$$
$$\mathbf{v} \longmapsto \mathbf{C}\mathbf{v}, \qquad\quad \mathbf{v} \longmapsto \mathbf{B}\mathbf{v}, \qquad\qquad\qquad \mathbf{v} \longmapsto \mathbf{A}\mathbf{v}.$$

This illustrates how we can prove results about matrices by using linear transformations. We can also prove results about linear transformations by using matrices.

It follows from the Composition Rule that $\mathbf{A}(\mathbf{B}\mathbf{C})$ is the matrix of the linear transformation $r \circ (s \circ t)$ and that $(\mathbf{A}\mathbf{B})\mathbf{C}$ is the matrix of the linear transformation $(r \circ s) \circ t$, with respect to the standard bases for the domain and codomain. The linear transformations $r \circ (s \circ t)$ and $(r \circ s) \circ t$ are equal, since $(r \circ (s \circ t))(\mathbf{v})$ and $((r \circ s) \circ t)(\mathbf{v})$ both mean $r(s(t(\mathbf{v})))$. It follows that $\mathbf{A}(\mathbf{B}\mathbf{C}) = (\mathbf{A}\mathbf{B})\mathbf{C}$. ∎

3.2 Invertible linear transformations

In this subsection we introduce the notion of an *invertible linear transformation*. Suppose that $t : V \longrightarrow W$ is a linear transformation which is one-one (no two elements of V have the same image) and is also onto (the image set $t(V)$ is the whole of W). Then t has an inverse function t^{-1} with domain W, such that

Inverse functions were introduced in Unit I2.

$$t^{-1}(t(\mathbf{v})) = \mathbf{v}, \quad \text{for each } \mathbf{v} \in V,$$

and

$$t(t^{-1}(\mathbf{w})) = \mathbf{w}, \quad \text{for each } \mathbf{w} \in W;$$

that is,

$$t^{-1} \circ t = i_V \quad \text{and} \quad t \circ t^{-1} = i_W.$$

We say that t is *invertible*.

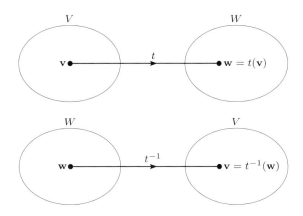

Definition The linear transformation $t : V \longrightarrow W$ is **invertible** if there exists an inverse function $t^{-1} : W \longrightarrow V$ such that

$$t^{-1} \circ t = i_V \quad \text{and} \quad t \circ t^{-1} = i_W.$$

Thus a linear transformation $t : V \longrightarrow W$ is invertible if and only if it is one-one and onto.

The linear transformation

$$t : \mathbb{R}^2 \longrightarrow \mathbb{R}^2$$
$$(x, y) \longmapsto (x, 0)$$

is not invertible, since it is not one-one; for example,

$$t(1, 1) = t(1, 2) = (1, 0).$$

The linear transformation

$$t : \mathbb{R}^2 \longrightarrow \mathbb{R}^3$$
$$(x, y) \longmapsto (x, y, 0)$$

is not invertible, since it is not onto; the image set $t(\mathbb{R}^2)$ is the (x, y)-plane, which is not the whole of the codomain $\mathbb{R}^3$.

Now consider the linear transformation

$$t : \mathbb{R}^2 \longrightarrow \mathbb{R}^2$$
$$(x, y) \longmapsto (2x, 2y).$$

We can check that t is one-one and onto and hence invertible, but what is the inverse function of t?

Use the techniques of Unit I2, Section 2.

Since t stretches each vector by a factor 2, we expect the inverse function of t to be the linear transformation

$$s : \mathbb{R}^2 \longrightarrow \mathbb{R}^2$$
$$(x, y) \longmapsto (\tfrac{1}{2}x, \tfrac{1}{2}y)$$

which contracts each vector by a factor 2. Since

$$s(t(x, y)) = s(2x, 2y) = (x, y)$$

and

$$t(s(x, y)) = t(\tfrac{1}{2}x, \tfrac{1}{2}y) = (x, y)$$

for each vector (x, y) in $\mathbb{R}^2$, $s \circ t$ and $t \circ s$ are both the identity transformation of $\mathbb{R}^2$, so s is the inverse function of t.

Exercise 3.4   Verify that the linear transformation

$$s : \mathbb{R}^2 \longrightarrow \mathbb{R}^2$$
$$(x, y) \longmapsto (x + y, 3x + 4y)$$

is the inverse function of the linear transformation

$$t : \mathbb{R}^2 \longrightarrow \mathbb{R}^2$$
$$(x, y) \longmapsto (4x - y, -3x + y).$$

In fact, the inverse of any linear transformation is a linear transformation. Unfortunately, it is not always obvious whether a given linear transformation $t : V \longrightarrow W$ is invertible. Even if we know that t is one-one and onto and hence invertible, it may not be clear what the inverse function of t is. If V and W are both *finite*-dimensional vector spaces, however, then t has a matrix representation. As we shall see, this can be used to determine whether t is invertible and, if so, to find the inverse function of t.

Theorem 3.2 Inverse Rule

Let $t : V \longrightarrow W$ be a linear transformation.

(a) If t is invertible, then $t^{-1} : W \longrightarrow V$ is also a linear transformation.

(b) If $\mathbf{A}$ is the matrix of t with respect to the bases E and F, then:

   (i) t is invertible if and only if $\mathbf{A}$ is invertible;

   (ii) if t is invertible, then $\mathbf{A}^{-1}$ is the matrix of t^{-1} with respect to the bases F and E.

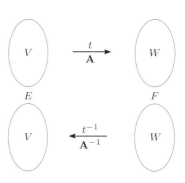

Proof   Let $t : V \longrightarrow W$ be a linear transformation.

(a) Suppose that t is invertible. We use Strategy 1.1 to show that the inverse function $t^{-1} : W \longrightarrow V$ is a linear transformation.

First we show that t^{-1} satisfies LT1:

$$t^{-1}(\mathbf{w}_1 + \mathbf{w}_2) = t^{-1}(\mathbf{w}_1) + t^{-1}(\mathbf{w}_2), \quad \text{for all } \mathbf{w}_1, \mathbf{w}_2 \in W.$$

Let $\mathbf{w}_1, \mathbf{w}_2 \in W$. Then, since t is invertible and hence onto, there exist $\mathbf{v}_1, \mathbf{v}_2 \in V$ such that $\mathbf{w}_1 = t(\mathbf{v}_1)$ and $\mathbf{w}_2 = t(\mathbf{v}_2)$. Since t satisfies LT1,

$$t^{-1}(\mathbf{w}_1 + \mathbf{w}_2) = t^{-1}(t(\mathbf{v}_1) + t(\mathbf{v}_2))$$
$$= t^{-1}(t(\mathbf{v}_1 + \mathbf{v}_2))$$
$$= \mathbf{v}_1 + \mathbf{v}_2.$$

Also,

$$t^{-1}(\mathbf{w}_1) + t^{-1}(\mathbf{w}_2) = t^{-1}(t(\mathbf{v}_1)) + t^{-1}(t(\mathbf{v}_2))$$
$$= \mathbf{v}_1 + \mathbf{v}_2,$$

so LT1 is satisfied.

Next we show that t^{-1} satisfies LT2:

$$t^{-1}(\alpha \mathbf{w}) = \alpha \, t^{-1}(\mathbf{w}), \quad \text{for all } \mathbf{w} \in W, \, \alpha \in \mathbb{R}.$$

Let $\mathbf{w} \in W$; then there exists $\mathbf{v} \in V$ such that $\mathbf{w} = t(\mathbf{v})$. Let $\alpha \in \mathbb{R}$; then, since t satisfies LT2,

$$t^{-1}(\alpha \mathbf{w}) = t^{-1}(\alpha \, t(\mathbf{v})) = t^{-1}(t(\alpha \mathbf{v})) = \alpha \mathbf{v}.$$

If you are short of time, omit this proof.

35

Also,

$$\alpha \, t^{-1}(\mathbf{w}) = \alpha \, t^{-1}(t(\mathbf{v})) = \alpha \mathbf{v},$$

so LT2 is satisfied.

Since LT1 and LT2 are satisfied, t^{-1} is a linear transformation.

(b) Let $\mathbf{A}$ be the matrix of t with respect to the bases E and F, so

$$t : \mathbf{v}_E \longmapsto \mathbf{A}\mathbf{v}_E = \mathbf{w}_F, \quad \text{for any vector } \mathbf{v} \in V.$$

First we show that if $\mathbf{A}$ is invertible, then t is invertible.

Suppose that $\mathbf{A}$ is invertible. Then we know that $\mathbf{A}$ is a square matrix and $\mathbf{A}^{-1}$ is also square, so we can define s to be the linear transformation with the matrix representation

Unit LA2, Section 4.

$$s : W \longrightarrow V$$
$$\mathbf{w}_F \longmapsto \mathbf{A}^{-1}\mathbf{w}_F = s(\mathbf{w})_E.$$

We show that s is the inverse function of t, and hence that t is invertible.

It follows from the Composition Rule that $s \circ t$ has the matrix representation

$$s \circ t : V \longrightarrow V$$
$$\mathbf{v}_E \longmapsto (\mathbf{A}^{-1}\mathbf{A})\mathbf{v}_E = \mathbf{I}\mathbf{v}_E = \mathbf{v}_E.$$

$\mathbf{I}$ denotes the identity matrix, which you met in Unit LA2, Section 3.

Thus $s(t(\mathbf{v})) = \mathbf{v}$ for each $\mathbf{v} \in V$; that is, $s \circ t = i_V$.

Similarly, it follows from the Composition Rule that $t \circ s$ has the matrix representation

$$t \circ s : W \longrightarrow W$$
$$\mathbf{w}_F \longmapsto (\mathbf{A}\mathbf{A}^{-1})\mathbf{w}_F = \mathbf{I}\mathbf{w}_F = \mathbf{w}_F.$$

Thus $t(s(\mathbf{w})) = \mathbf{w}$ for each $\mathbf{w} \in E$; that is, $t \circ s = i_W$.

Since $s \circ t = i_V$ and $t \circ s = i_W$, it follows that s is the inverse function of t, so t is invertible.

We now show that if t is invertible, then $\mathbf{A}$ is invertible and $\mathbf{A}^{-1}$ is the matrix of t^{-1} with respect to the bases F and E.

Suppose that t is invertible. Since t^{-1} is a linear transformation, it has a matrix representation

Theorem 2.1

$$t^{-1} : W \longrightarrow V$$
$$\mathbf{w}_F \longmapsto \mathbf{B}\mathbf{w}_F = t^{-1}(\mathbf{w})_E.$$

We show that $\mathbf{B} = \mathbf{A}^{-1}$.

It follows from the Composition Rule that $t^{-1} \circ t$ has the matrix representation

$$t^{-1} \circ t : V \longrightarrow V$$
$$\mathbf{v}_E \longmapsto (\mathbf{B}\mathbf{A})\mathbf{v}_E.$$

Since $(t^{-1} \circ t)(\mathbf{v}) = \mathbf{v}$ for each $\mathbf{v} \in V$, it follows that

$$(\mathbf{B}\mathbf{A})\mathbf{v}_E = \mathbf{v}_E \quad \text{for all } \mathbf{v} \in V.$$

Thus $\mathbf{B}\mathbf{A} = \mathbf{I}$.

Similarly, it follows from the Composition Rule that $t \circ t^{-1}$ has the matrix representation

$$t \circ t^{-1} : W \longrightarrow W$$
$$\mathbf{w}_F \longmapsto (\mathbf{AB})\mathbf{w}_F.$$

Since $(t \circ t^{-1})(\mathbf{w}) = \mathbf{w}$ for each $\mathbf{w} \in W$, it follows that

$$(\mathbf{AB})\mathbf{w}_F = \mathbf{w}_F \quad \text{for all } \mathbf{w} \in W.$$

Thus $\mathbf{AB} = \mathbf{I}$.

Since

$$\mathbf{BA} = \mathbf{AB} = \mathbf{I},$$

it follows that $\mathbf{A}$ is invertible and $\mathbf{B} = \mathbf{A}^{-1}$.

Unit LA2, Section 4.

This completes the proof.   ■

One consequence of the Inverse Rule is that if $t : V \longrightarrow W$ is an invertible linear transformation, then any matrix of t must be invertible and hence square. Since a matrix of t has m rows and n columns, where m is the dimension of W and n is the dimension of V, it follows that $m = n$; that is, the vector spaces V and W must have the same dimension.

Unit LA2, Section 4.

Corollary to Theorem 3.2

Let $t : V \longrightarrow W$ be an invertible linear transformation, where V and W are finite dimensional. Then

$$\dim V = \dim W.$$

Remark   It follows that if $t : V \longrightarrow W$ is a linear transformation and V and W have *different* finite dimensions, then t is not invertible. For example, the linear transformation

$$t : \mathbb{R}^3 \longrightarrow \mathbb{R}^2$$
$$(x, y, z) \longmapsto (2x + y, x - y)$$

is not invertible, since the domain and codomain have different dimensions.

Now suppose that $t : V \longrightarrow W$ is a linear transformation and that V and W have the *same* finite dimension. It follows from the Inverse Rule that you can use the following strategy to determine whether or not t is invertible.

Strategy 3.1   To determine whether or not a linear transformation $t : V \longrightarrow W$ is invertible, where V and W are n-dimensional vector spaces with bases E and F, respectively.

1. Find a matrix representation of t,

$$\mathbf{v}_E \longmapsto \mathbf{A}\mathbf{v}_E = t(\mathbf{v})_F.$$

2. Evaluate $\det \mathbf{A}$.

If $\det \mathbf{A} = 0$, then t is not invertible.

If $\det \mathbf{A} \neq 0$, then t is invertible and $t^{-1} : W \longrightarrow V$ has the matrix representation

$$\mathbf{w}_F \longmapsto \mathbf{A}^{-1}\mathbf{w}_F = t^{-1}(\mathbf{w})_E.$$

You saw that a matrix is invertible if and only if its determinant is non-zero in Unit LA2, Section 5.

Example 3.1   Show that the following linear transformation t is invertible and find the inverse function of t.

$$t : \mathbb{R}^2 \longrightarrow \mathbb{R}^2$$
$$(x, y) \longmapsto (x + y, 2y)$$

Solution   Following Strategy 3.1, we first find a matrix representation of t.

Since

$$t(1, 0) = (1, 0) \quad \text{and} \quad t(0, 1) = (1, 2),$$

the matrix representation of t with respect to the standard basis for the domain and codomain is

We use Strategy 2.1.

$$\begin{pmatrix} x \\ y \end{pmatrix} \longmapsto \begin{pmatrix} 1 & 1 \\ 0 & 2 \end{pmatrix} \begin{pmatrix} x \\ y \end{pmatrix} = \begin{pmatrix} x + y \\ 2y \end{pmatrix}.$$

The next step is to evaluate the determinant of the matrix

$$\mathbf{A} = \begin{pmatrix} 1 & 1 \\ 0 & 2 \end{pmatrix}.$$

We have

$$\det \mathbf{A} = \begin{vmatrix} 1 & 1 \\ 0 & 2 \end{vmatrix} = (1 \times 2) - (1 \times 0) = 2.$$

Since $\det \mathbf{A}$ is non-zero, t is invertible.

We now find the inverse function of t. According to Strategy 3.1, $t^{-1} : \mathbb{R}^2 \longrightarrow \mathbb{R}^2$ has the matrix representation

$$\mathbf{v} \longmapsto \mathbf{A}^{-1} \mathbf{v}$$

with respect to the standard basis for the domain and codomain. Since

$$\mathbf{A}^{-1} = \tfrac{1}{2} \begin{pmatrix} 2 & -1 \\ 0 & 1 \end{pmatrix} = \begin{pmatrix} 1 & -\tfrac{1}{2} \\ 0 & \tfrac{1}{2} \end{pmatrix},$$

See Unit LA2, Strategy 5.1.

it follows that t^{-1} has the matrix representation

$$\begin{pmatrix} x \\ y \end{pmatrix} \longmapsto \begin{pmatrix} 1 & -\tfrac{1}{2} \\ 0 & \tfrac{1}{2} \end{pmatrix} \begin{pmatrix} x \\ y \end{pmatrix} = \begin{pmatrix} x - \tfrac{1}{2}y \\ \tfrac{1}{2}y \end{pmatrix}.$$

So t^{-1} is the linear transformation

$$t^{-1} : \mathbb{R}^2 \longrightarrow \mathbb{R}^2$$
$$(x, y) \longmapsto \left(x - \tfrac{1}{2}y, \tfrac{1}{2}y \right). \quad \blacksquare$$

Exercise 3.5   Determine which of the following linear transformations are invertible. Find the inverse function of each invertible linear transformation.

(a) $t : \mathbb{R}^2 \longrightarrow \mathbb{R}^2$
$(x, y) \longmapsto (2x + y, 4x + 2y)$

(b) $t : \mathbb{R}^2 \longrightarrow \mathbb{R}^2$
$(x, y) \longmapsto (x - y, 3x + y)$

(c) $t : \mathbb{R}^3 \longrightarrow \mathbb{R}^3$
$(x, y, z) \longmapsto (2x, 3y - x, z)$

(d) $t : P_3 \longrightarrow P_2$
$p(x) \longmapsto p'(x)$

In Example 3.1 we considered the linear transformation

$$t : \mathbb{R}^2 \longrightarrow \mathbb{R}^2$$
$$(x, y) \longmapsto (x + y, 2y).$$

We found the matrix $\mathbf{A}$ of t with respect to the *standard* basis for $\mathbb{R}^2$, and showed that $\det \mathbf{A} = 2$. In fact, whatever bases we had chosen for the domain and codomain, we would still have obtained a matrix of t with determinant equal to 2.

We omit the proof of this result.

It can be shown that the magnitude of the determinant of a matrix of t is the 'stretching factor' of t. Since $\det \mathbf{A} = 2$ in the above case, areas are doubled under t.

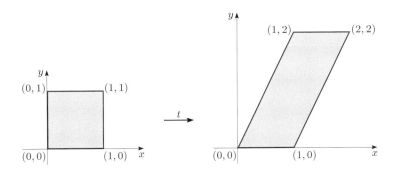

For a linear transformation $t : \mathbb{R}^2 \longrightarrow \mathbb{R}^2$ and a matrix $\mathbf{A}$ of t with $\det \mathbf{A} = 0$, the image of a unit square under t is a line or a point—these have zero area. So, in this case, t is not invertible.

3.3 Isomorphisms

We have seen that there are invertible linear transformations from $\mathbb{R}^2$ to itself. In fact, whenever the vector spaces V and W have the *same* finite dimension, we can construct an invertible linear transformation from V to W.

For example, consider the two-dimensional vector spaces $\mathbb{R}^2$ and P_2. The linear transformation

$$t : P_2 \longrightarrow \mathbb{R}^2$$
$$a + bx \longmapsto (a, b)$$

is one-one and onto and hence invertible. By looking at a matrix representation of t, we can see how to construct an invertible linear transformation from V to W, whenever V and W have the same finite dimension.

For t above, take the standard bases $E = \{1, x\}$ for P_2 and $F = \{(1, 0), (0, 1)\}$ for $\mathbb{R}^2$. Then $t(1) = (1, 0)$ and $t(x) = (0, 1)$, so t has the matrix representation

$$\begin{pmatrix} a \\ b \end{pmatrix}_E \longmapsto \begin{pmatrix} 1 & 0 \\ 0 & 1 \end{pmatrix} \begin{pmatrix} a \\ b \end{pmatrix}_E = \begin{pmatrix} a \\ b \end{pmatrix}_F,$$

that is,

$$\mathbf{v}_E \longmapsto \mathbf{I}_2 \mathbf{v}_E = \mathbf{w}_F.$$

More generally, let V and W be n-dimensional vector spaces, let E be a basis for V and let F be a basis for W. Then

$$t : V \longrightarrow W$$
$$\mathbf{v}_E \longmapsto \mathbf{I}_n \mathbf{v}_E = \mathbf{w}_F$$

is a linear transformation from V to W.

Since the identity matrix $\mathbf{I}_n$ is invertible, it follows from the Inverse Rule that t is invertible. Note that t maps the first basis vector in E to the first basis vector in F, the second basis vector in E to the second basis vector in F, and so on. We say that t is an *isomorphism* from V to W.

Definitions   The vector spaces V and W are **isomorphic** if there exists an invertible linear transformation $t : V \longrightarrow W$. Such a function t is an **isomorphism**.

Exercise 3.6   Write down an isomorphism from P_3 to $\mathbb{R}^3$.

Suppose that V and W are finite-dimensional vector spaces. We have just seen that if $\dim V = \dim W$, then there is an invertible linear transformation $t : V \longrightarrow W$; that is, V and W are isomorphic.

In particular, each n-dimensional vector space is isomorphic to $\mathbb{R}^n$.

We also know that if V and W have *different* dimensions, then there are *no* invertible linear transformations from V to W; that is, V and W are not isomorphic. Thus we have proved the following result.

This is the remark following the corollary to Theorem 3.1.

Theorem 3.3   The finite-dimensional vector spaces V and W are isomorphic if and only if

$$\dim V = \dim W.$$

Exercise 3.7   State which of the following vector spaces are isomorphic to each other:

$$\mathbb{R}^2, \quad \mathbb{R}^3, \quad \mathbb{C}, \quad P_2, \quad P_3.$$

Further exercises

Exercise 3.8   Let t and s be the linear transformations

$$t : \mathbb{R}^2 \longrightarrow \mathbb{R}^2 \qquad \text{and} \qquad s : \mathbb{R}^2 \longrightarrow \mathbb{R}^2$$
$$(x, y) \longmapsto (x + y, 2y) \qquad (x, y) \longmapsto (3x, 2y).$$

Determine the following linear transformations.

(a) $s \circ t$ (b) $t \circ s$ (c) $t \circ t$

Exercise 3.9   Let t and s be the linear transformations

$$t : \mathbb{R}^3 \longrightarrow \mathbb{R}^2 \qquad \text{and} \qquad s : \mathbb{R}^2 \longrightarrow \mathbb{R}^2$$
$$(x, y, z) \longmapsto (x + y, x - z) \qquad (x, y) \longmapsto (3x + y, 2x - y).$$

Use the Composition Rule and the solution to Exercise 2.8 to find the matrix representation of the linear transformation $s \circ t$ with respect to the standard bases for the domain and codomain.

Exercise 3.10   Let $V = \{f(x) : f(x) = ae^x \cos x + be^x \sin x, \ a, b \in \mathbb{R}\}$ and let t be the linear transformation

$$t : V \longrightarrow V$$
$$f(x) \longmapsto f'(x).$$

Use the Composition Rule and the solution to Exercise 2.10 to find the matrix representation of the linear transformation

$$t \circ t : V \longrightarrow V$$
$$f(x) \longmapsto f''(x)$$

with respect to the basis $E = \{e^x \cos x, e^x \sin x\}$ for the domain and codomain.

Hence find the image of a function $f(x) = ae^x \cos x + be^x \sin x$ under $t \circ t$.

Exercise 3.11   Determine whether or not each of the following linear transformations is invertible. Find the inverse of each invertible linear transformation.

(a) $t : \mathbb{R}^2 \longrightarrow \mathbb{R}^2$

$\quad (x, y) \longmapsto (3x + y, y)$

(b) $t : \mathbb{R}^2 \longrightarrow \mathbb{R}^2$

$\quad (x, y) \longmapsto (2x + 6y, x + 3y)$

(c) $t : \mathbb{R}^2 \longrightarrow \mathbb{R}^4$

$\quad (x, y) \longmapsto (x, y, x, y)$

(d) $\quad t : \mathbb{R}^3 \longrightarrow \mathbb{R}^3$

$\quad (x, y, z) \longmapsto (x + z, 2x + y + 3z, 2y + z)$

Exercise 3.12   Use the solution to Exercise 2.11(a) to determine whether or not the following linear transformation is invertible.

$$t : P_3 \longrightarrow P_3$$
$$p(x) \longmapsto p(x) + p'(x)$$

Exercise 3.13   Let $V = \{f(x) : f(x) = ae^x \cos x + be^x \sin x, \ a, b \in \mathbb{R}\}$. Write down an isomorphism:

(a) from V to $\mathbb{R}^2$;   (b) from V to P_2.

Exercise 3.14   (Harder) Prove that the set of invertible linear transformations from $\mathbb{R}^n$ to itself forms a group under composition.

4   Image and kernel

After working through this section, you should be able to:

(a) explain the meaning of the terms *image* and *kernel* of a linear transformation;

(b) find a basis for the image of a given linear transformation;

(c) find the kernel of a given linear transformation;

(d) understand the relationship between the dimension of the image and the dimension of the kernel of a linear transformation.

In the previous section you saw that a linear transformation $t : V \longrightarrow W$ is invertible if t is one-one and onto; that is, each element of W is the image of exactly one element of V.

In this section we give a strategy for determining which elements of W are the images of elements of V. We then investigate conditions under which an element of W is the image of more than one element of V. This enables us to prove an important result known as the *Dimension Theorem*.

Finally, we use the Dimension Theorem to show how the number of possible solutions to a system of m simultaneous linear equations in n unknowns depends on the values of m and n.

4.1   Image of a linear transformation

Let t be the linear transformation

$$t : \mathbb{R}^3 \longrightarrow \mathbb{R}^3$$

$$(x, y, z) \longmapsto (x, y, 0).$$

This projects each vector (x, y, z) onto the vector $(x, y, 0)$ in the (x, y)-plane.

A vector $\mathbf{w}$ in the codomain $\mathbb{R}^3$ is the image of a vector $\mathbf{v}$ in the domain $\mathbb{R}^3$ if and only if $\mathbf{w}$ is in the (x, y)-plane. We say that the (x, y)-plane is the *image* of t. This is a two-dimensional subspace of the codomain $\mathbb{R}^3$.

In general, the image of a function $t : V \longrightarrow W$ is the set of all the elements of W that are images of elements of V.

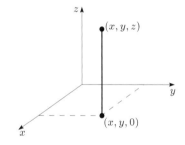

We introduced the image of a function in Unit I2.

Definition   The **image** of a linear transformation $t : V \longrightarrow W$ is the set

$$\mathrm{Im}(t) = \{\mathbf{w} \in W : \mathbf{w} = t(\mathbf{v}), \text{ for some } \mathbf{v} \in V\}.$$

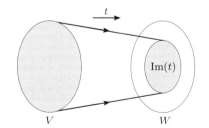

Exercise 4.1   Give a geometric description of the image of each of the following linear transformations. In each case, state whether the image is a subspace of the codomain.

(a) $t : \mathbb{R}^3 \longrightarrow \mathbb{R}^2$

$(x, y, z) \longmapsto (x, 0)$

(b) $t : \mathbb{R}^2 \longrightarrow \mathbb{R}^2$

$(x, y) \longmapsto (x, x)$

For each of the linear transformations in Exercise 4.1, the image is a subspace of the codomain. This is true for all linear transformations.

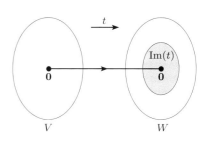

> **Theorem 4.1** Let $t : V \longrightarrow W$ be a linear transformation. Then $\mathrm{Im}(t)$ is a subspace of the codomain W.

Proof   We follow Strategy 4.1 in Unit LA3.

We check first that $\mathbf{0} \in \mathrm{Im}(t)$.

Since t is a linear transformation, $t(\mathbf{0}) = \mathbf{0}$, so $\mathbf{0} \in \mathrm{Im}(t)$.

We check next that $\mathrm{Im}(t)$ is closed under vector addition.

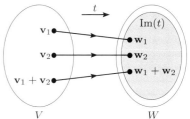

Let $\mathbf{w}_1, \mathbf{w}_2 \in \mathrm{Im}(t)$. Then there exist vectors $\mathbf{v}_1, \mathbf{v}_2 \in V$ such that $\mathbf{w}_1 = t(\mathbf{v}_1)$ and $\mathbf{w}_2 = t(\mathbf{v}_2)$. Since t is a linear transformation,

$$\mathbf{w}_1 + \mathbf{w}_2 = t(\mathbf{v}_1) + t(\mathbf{v}_2) = t(\mathbf{v}_1 + \mathbf{v}_2).$$

Since V is closed under vector addition, $\mathbf{v}_1 + \mathbf{v}_2 \in V$, so $\mathbf{w}_1 + \mathbf{w}_2 \in \mathrm{Im}(t)$.

Finally, we show that $\mathrm{Im}(t)$ is closed under scalar multiplication.

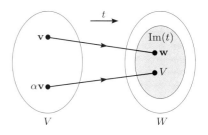

Let $\mathbf{w} \in \mathrm{Im}(t)$ and $\alpha \in \mathbb{R}$. Then there exists $\mathbf{v} \in V$ such that $\mathbf{w} = t(\mathbf{v})$ and, since t is a linear transformation,

$$\alpha \mathbf{w} = \alpha\, t(\mathbf{v}) = t(\alpha \mathbf{v}).$$

Since V is closed under scalar multiplication, $\alpha \mathbf{v} \in V$, so $\alpha \mathbf{w} \in \mathrm{Im}(t)$.

Thus $\mathrm{Im}(t)$ is a subspace of W.   ∎

For the linear transformations studied so far, it has been easy to write down their images. In general, this is not the case, so we need some way of determining the image of a linear transformation.

If we know the image of each vector in a basis for V, then we can find the image of each vector in V. Thus the image of t is determined by the images of the domain basis vectors.

This follows from Theorem 1.3.

For example, consider the linear transformation

$$t : \mathbb{R}^3 \longrightarrow \mathbb{R}^3$$
$$(x, y, z) \longmapsto (x, y, 0).$$

The standard basis for the domain $\mathbb{R}^3$ is $\{(1,0,0), (0,1,0), (0,0,1)\}$.

The images of the vectors in this basis are

$$(1,0,0), \quad (0,1,0), \quad (0,0,0).$$

These vectors all lie in $\mathrm{Im}(t)$, which is the (x, y)-plane, and they *span* $\mathrm{Im}(t)$; that is, each vector in $\mathrm{Im}(t)$ can be written as a linear combination of the vectors $(1,0,0)$, $(0,1,0)$ and $(0,0,0)$.

> **Exercise 4.2**   Let t be the linear transformation
>
> $$t : \mathbb{R}^2 \longrightarrow \mathbb{R}^2$$
> $$(x, y) \longmapsto (x, x).$$
>
> Determine the images of the vectors in the standard basis $\{(1,0), (0,1)\}$ for the domain $\mathbb{R}^2$. Do these image vectors span $\mathrm{Im}(t)$?

You found the image of this linear transformation in Exercise 4.1(b).

We show now that, for any linear transformation, the images of the domain basis vectors span the image.

Let $t : V \longrightarrow W$ be a linear transformation and let $\{\mathbf{e}_1, \ldots, \mathbf{e}_n\}$ be a basis for V. If $\mathbf{w} \in \operatorname{Im}(t)$, then $\mathbf{w} = t(\mathbf{v})$ for some $\mathbf{v}$ in V. Since $\{\mathbf{e}_1, \ldots, \mathbf{e}_n\}$ is a basis for V, there exist real numbers $v_1, \ldots, v_n$ such that

$$\mathbf{v} = v_1 \mathbf{e}_1 + \cdots + v_n \mathbf{e}_n.$$

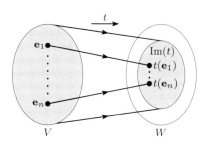

Since t is a linear transformation, it follows that

$$\mathbf{w} = t(\mathbf{v}) = v_1 t(\mathbf{e}_1) + \cdots + v_n t(\mathbf{e}_n).$$

Thus $\mathbf{w}$ is a linear combination of the vectors $t(\mathbf{e}_1), \ldots, t(\mathbf{e}_n)$.

See Theorem 1.3.

So $\{t(\mathbf{e}_1), \ldots, t(\mathbf{e}_n)\}$ is a spanning set for $\operatorname{Im}(t)$, as claimed.

Since a basis is a linearly independent spanning set, we now give a strategy that enables us to find a basis for the image of a linear transformation.

Strategy 4.1   To find a basis for $\operatorname{Im}(t)$, where $t : V \longrightarrow W$ is a linear transformation.

1. Find a basis $\{\mathbf{e}_1, \ldots, \mathbf{e}_n\}$ for the domain V.

2. Determine the vectors $t(\mathbf{e}_1), \ldots, t(\mathbf{e}_n)$.

3. If there is a vector $\mathbf{v}$ in $S = \{t(\mathbf{e}_1), \ldots, t(\mathbf{e}_n)\}$ that is a linear combination of the other vectors in S, then discard $\mathbf{v}$ to give the set $S_1 = S - \{\mathbf{v}\}$.

4. If there is a vector $\mathbf{v}_1$ in S_1 such that $\mathbf{v}_1$ is a linear combination of the other vectors in S_1, then discard $\mathbf{v}_1$ to give the set $S_2 = S_1 - \{\mathbf{v}_1\}$.

Continue discarding vectors in this way until you obtain a linearly independent set. This set is a basis for $\operatorname{Im}(t)$.

S_1 spans V by Unit LA3, Theorem 3.1.

Once we know a basis for the image of a linear transformation, we know everything that we need to know about the image; in particular, we know its dimension.

Example 4.1   Let t be the linear transformation

$$t : \mathbb{R}^3 \longrightarrow \mathbb{R}^2$$

$$(x, y, z) \longmapsto (x + 2y + 3z, 4x + y - 2z).$$

Find a basis for $\operatorname{Im}(t)$ and state the dimension of $\operatorname{Im}(t)$.

Solution   We follow the steps of Strategy 4.1.

We take the standard basis $\{(1, 0, 0), (0, 1, 0), (0, 0, 1)\}$ for the domain $\mathbb{R}^3$.

We determine the images of these basis vectors:

$$t(1, 0, 0) = (1, 4), \quad t(0, 1, 0) = (2, 1), \quad t(0, 0, 1) = (3, -2).$$

The set $\{(1, 4), (2, 1), (3, -2)\}$ is not linearly independent. In fact,

$$(3, -2) = 2(2, 1) - (1, 4),$$

so we can discard $(3, -2)$ to give the set $\{(1, 4), (2, 1)\}$.

We could have discarded $(1, 4)$ or $(2, 1)$ instead.

The vectors $(1, 4)$ and $(2, 1)$ are linearly independent, so $\{(1, 4), (2, 1)\}$ is a basis for $\operatorname{Im}(t)$.

Since the basis has two elements, $\operatorname{Im}(t)$ is a two-dimensional subspace of the codomain $\mathbb{R}^2$. ∎

Exercise 4.3 For each of the following linear transformations t, find a basis for $\operatorname{Im}(t)$ and state the dimension of $\operatorname{Im}(t)$.

(a) $t : \mathbb{R}^2 \longrightarrow \mathbb{R}^2$

   $(x, y) \longmapsto (x, 2x + y)$

(b) $t : \mathbb{R}^3 \longrightarrow \mathbb{R}^3$

   $(x, y, z) \longmapsto (x + 2y + 3z, x + z, x + y + 2z)$

(c) $t : P_3 \longrightarrow P_2$

   $p(x) \longmapsto p'(x)$

For the linear transformation t in Exercise 4.3(b), $\operatorname{Im}(t)$ is a two-dimensional subspace of the codomain $\mathbb{R}^3$. Thus $\operatorname{Im}(t)$ is a plane through the origin with equation

$$ax + by + cz = 0,$$

for some $a, b, c \in \mathbb{R}$ not all zero. It is possible to use the basis that you found for $\operatorname{Im}(t)$ in Exercise 4.3(b) to work out the values of a, b and c.

For example, $\{(1, 1, 1), (2, 0, 1)\}$ is a basis for $\operatorname{Im}(t)$. Since the basis vectors belong to $\operatorname{Im}(t)$, the values a, b and c satisfy the simultaneous equations

$$\begin{cases} a + b + c = 0, \\ 2a \quad\;\; + c = 0. \end{cases}$$

The second equation gives $c = -2a$. Substituting this into the first equation gives $b = a$. So $\operatorname{Im}(t)$ is the plane with equation

$$ax + ay - 2az = 0$$

or, equivalently,

$$x + y - 2z = 0.$$

Finally we note that a linear transformation $t : V \longrightarrow W$ is onto when every element of W is the image of an element of V. Thus we have the following result.

Property 4.1 A linear transformation t is onto if and only if $\operatorname{Im}(t) = W$.

Exercise 4.4 Which of the linear transformations in Exercise 4.3 are onto?

4.2 Kernel of a linear transformation

We have seen how to find the image of a linear transformation $t : V \longrightarrow W$. Now suppose that $\mathbf{w}$ belongs to the image of t. How can we find all the vectors in V that map to $\mathbf{w}$?

We begin by looking at the case when $\mathbf{w}$ is the zero vector. We know that $t(\mathbf{0}) = \mathbf{0}$, but it is possible that there are also some non-zero vectors in V that are mapped to $\mathbf{0}$.

For example, let t be the linear transformation

$$t : \mathbb{R}^3 \longrightarrow \mathbb{R}^3$$

$$(x, y, z) \longmapsto (x, y, 0).$$

Then $t(x, y, z) = \mathbf{0}$ if and only if $(x, y, 0) = (0, 0, 0)$; that is, if and only if $x = 0$ and $y = 0$.

Thus the set of vectors that are mapped to $\mathbf{0}$ is the whole of the z-axis. This set is a one-dimensional subspace of the domain $\mathbb{R}^3$. We call this set the *kernel* of t.

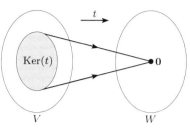

Definition   The **kernel** of a linear transformation $t : V \longrightarrow W$ is the set

$$\mathrm{Ker}(t) = \{\mathbf{v} \in V : t(\mathbf{v}) = \mathbf{0}\}.$$

Exercise 4.5   Give a geometric description of the kernel of each of the following linear transformations. In each case, state whether the kernel is a subspace of the domain.

(a) $t : \mathbb{R}^3 \longrightarrow \mathbb{R}^2$
       $(x, y, z) \longmapsto (x, 0)$

(b) $t : \mathbb{R}^2 \longrightarrow \mathbb{R}^2$
       $(x, y) \longmapsto (x, x)$

For each of the linear transformations in Exercise 4.5, the kernel is a subspace of the domain. This is true for all linear transformations.

Theorem 4.2   Let $t : V \longrightarrow W$ be a linear transformation. Then $\mathrm{Ker}(t)$ is a subspace of the domain V.

Proof   We follow Strategy 4.1 in Unit LA3.

First we show that $\mathbf{0} \in \mathrm{Ker}(t)$.

Since t is a linear transformation, $t(\mathbf{0}) = \mathbf{0}$, so $\mathbf{0} \in \mathrm{Ker}(t)$.

Next we show that $\mathrm{Ker}(t)$ is closed under vector addition.

Let $\mathbf{v}_1, \mathbf{v}_2 \in \mathrm{Ker}(t)$. Since t is a linear transformation,

$$t(\mathbf{v}_1 + \mathbf{v}_2) = t(\mathbf{v}_1) + t(\mathbf{v}_2) = \mathbf{0} + \mathbf{0} = \mathbf{0},$$

so $\mathbf{v}_1 + \mathbf{v}_2 \in \mathrm{Ker}(t)$, as required.

Finally, we show that $\mathrm{Ker}(t)$ is closed under scalar multiplication.

Let $\mathbf{v} \in \mathrm{Ker}(t)$ and $\alpha \in \mathbb{R}$. Since t is a linear transformation,

$$t(\alpha\mathbf{v}) = \alpha\, t(\mathbf{v}) = \alpha\mathbf{0} = \mathbf{0},$$

so $\alpha\mathbf{v} \in \mathrm{Ker}(t)$.

Thus $\mathrm{Ker}(t)$ is a subspace of V.   ∎

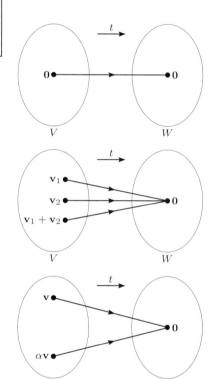

When finding the kernel of a linear transformation, we often need to solve a system of simultaneous linear equations. This sometimes involves using row-reduction.

Example 4.2   Let t be the linear transformation

$$t : \mathbb{R}^3 \longrightarrow \mathbb{R}^2$$
$$(x, y, z) \longmapsto (x + 2y + 3z, 4x + y - 2z).$$

Find the kernel of t and the dimension of the kernel.

Solution   We wish to find the set of vectors (x, y, z) in $\mathbb{R}^3$ that satisfy

$$t(x, y, z) = \mathbf{0},$$

that is,

$$(x + 2y + 3z, 4x + y - 2z) = (0, 0),$$

so we solve the simultaneous equations

$$\begin{cases} x + 2y + 3z = 0, \\ 4x + y - 2z = 0. \end{cases}$$

To solve these equations using row-reduction, we use the augmented matrix.

Unit LA2, Section 2.

$$
\begin{matrix} \mathbf{r}_1 \\ \mathbf{r}_2 \end{matrix}
\qquad
\left(\begin{array}{ccc|c} 1 & 2 & 3 & 0 \\ 4 & 1 & -2 & 0 \end{array} \right)
$$

$$
\mathbf{r}_2 \to \mathbf{r}_2 - 4\mathbf{r}_1
\qquad
\left(\begin{array}{ccc|c} 1 & 2 & 3 & 0 \\ 0 & -7 & -14 & 0 \end{array} \right)
$$

$$
\mathbf{r}_2 \to -\tfrac{1}{7}\mathbf{r}_2
\qquad
\left(\begin{array}{ccc|c} 1 & 2 & 3 & 0 \\ 0 & 1 & 2 & 0 \end{array} \right)
$$

$$
\mathbf{r}_1 \to \mathbf{r}_1 - 2\mathbf{r}_2
\qquad
\left(\begin{array}{ccc|c} 1 & 0 & -1 & 0 \\ 0 & 1 & 2 & 0 \end{array} \right)
$$

The augmented matrix is now in row-reduced form. It gives

$$\begin{cases} x \quad\;\; - \;\; z = 0, \\ \quad\;\; y + 2z = 0. \end{cases}$$

Assigning the parameter k to the unknown z, we obtain

$$x = k, \quad y = -2k, \quad z = k.$$

So the kernel of t is

$$\mathrm{Ker}(t) = \{(k, -2k, k) : k \in \mathbb{R}\},$$

that is, $\mathrm{Ker}(t)$ is the line through $(0, 0, 0)$ and $(1, -2, 1)$.

Thus $\mathrm{Ker}(t)$ is a one-dimensional subspace of the domain $\mathbb{R}^3$. ■

Exercise 4.6   For each of the following linear transformations t, find the kernel of t and the dimension of the kernel.

(a) $t : \mathbb{R}^2 \longrightarrow \mathbb{R}^2$

$\qquad (x, y) \longmapsto (x, 2x + y)$

(b) $t : \mathbb{R}^3 \longrightarrow \mathbb{R}^3$

$\qquad (x, y, z) \longmapsto (x + 2y + 3z, x + z, x + y + 2z)$

We now look at an example involving vector spaces of polynomials.

Example 4.3   Find the kernel of the linear transformation

$t : P_3 \longrightarrow P_3$

$p(x) \longmapsto p(x) + p(2).$

Solution   Let $p(x) = a + bx + cx^2$ be a polynomial in P_3. Then

$$t(p(x)) = a + bx + cx^2 + a + 2b + 4c$$
$$= 2a + 2b + 4c + bx + cx^2.$$

We wish to find those polynomials in P_3 that satisfy $t(p(x)) = \mathbf{0}$; that is,

$$2a + 2b + 4c + bx + cx^2 = 0, \quad \text{for all } x \in \mathbb{R},$$

so we solve the simultaneous equations

$$\begin{cases} 2a + 2b + 4c = 0, \\ \quad\quad b \quad\quad = 0, \\ \quad\quad\quad\quad c = 0. \end{cases}$$

Putting $b = 0$ and $c = 0$ into the first equation gives $a = 0$. So the simultaneous equations are satisfied only when $a = 0$, $b = 0$ and $c = 0$.

Thus the only polynomial in the kernel of t is the zero polynomial $p(x) = 0$; that is,

$$\text{Ker}(t) = \{\mathbf{0}\}. \quad \blacksquare$$

Exercise 4.7   Find the kernel of the linear transformation

$$t : P_3 \longrightarrow P_2$$
$$p(x) \longmapsto p'(x).$$

Find the dimension of the kernel.

For a given linear transformation $t : V \longrightarrow W$, we know how to find all the vectors in V that map to $\mathbf{0}$ in W. Now suppose that $\mathbf{b}(\neq \mathbf{0})$ is some particular vector in W. How do we find all the vectors in V that map to $\mathbf{b}$? There is a close relationship between the vectors that map to $\mathbf{b}$ and those that map to $\mathbf{0}$: if we know *one* vector $\mathbf{a}$ in V that maps to $\mathbf{b}$, that is, $t(\mathbf{a}) = \mathbf{b}$, then *every* vector $\mathbf{x}$ in V that maps to $\mathbf{b}$ may be written in the form $\mathbf{x} = \mathbf{a} + \mathbf{k}$, where $t(\mathbf{k}) = \mathbf{0}$, that is, $\mathbf{k} \in \text{Ker}(t)$. We state this formally in the following theorem.

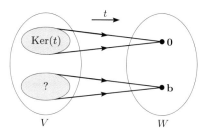

Theorem 4.3   Solution Set Theorem

Let $t : V \longrightarrow W$ be a linear transformation. Let $\mathbf{b} \in W$ and let $\mathbf{a}$ be one vector in V that maps to $\mathbf{b}$, that is, $t(\mathbf{a}) = \mathbf{b}$. Then the solution set of the equation $t(\mathbf{x}) = \mathbf{b}$ is

$$\{\mathbf{x} : \mathbf{x} = \mathbf{a} + \mathbf{k} \text{ for some } \mathbf{k} \in \text{Ker}(t)\}.$$

Proof   The proof is in two parts.

First we show that each vector $\mathbf{x}$ of the given form is a solution of $t(\mathbf{x}) = \mathbf{b}$. Let $\mathbf{x} = \mathbf{a} + \mathbf{k}$, where $\mathbf{k} \in \text{Ker}(t)$. Then

$$t(\mathbf{x}) = t(\mathbf{a} + \mathbf{k}) = t(\mathbf{a}) + t(\mathbf{k}) = \mathbf{b} + \mathbf{0} = \mathbf{b}.$$

Conversely, we show that each vector $\mathbf{x}$ in the solution set has the given form. Let $t(\mathbf{x}) = \mathbf{b}$, where $\mathbf{x} \in V$. Then

$$t(\mathbf{x} - \mathbf{a}) = t(\mathbf{x}) - t(\mathbf{a}) = \mathbf{b} - \mathbf{b} = \mathbf{0},$$

so $\mathbf{x} - \mathbf{a} \in \text{Ker}(t)$; that is, $\mathbf{x} = \mathbf{a} + \mathbf{k}$, for some $\mathbf{k} \in \text{Ker}(t)$. $\quad \blacksquare$

Finally, we recall that a linear transformation $t : V \longrightarrow W$ is one-one if and only if no two elements in V have the same image. Thus we have the following result.

Property 4.2   A linear transformation t is one-one if and only if $\mathrm{Ker}(t) = \{\mathbf{0}\}$.

Exercise 4.8   Which of the linear transformations in Exercises 4.6 and 4.7 are one-one?

4.3 Dimension Theorem

We have seen that a linear transformation $t : V \longrightarrow W$ has two particular subspaces associated with it: $\mathrm{Ker}(t)$ in the domain V and $\mathrm{Im}(t)$ in the codomain W.

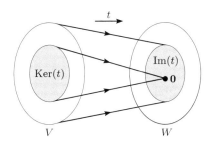

There is a remarkable connection between the dimensions of these two subspaces and the dimension of the domain V.

Let t be the linear transformation

$$t : \mathbb{R}^3 \longrightarrow \mathbb{R}^3$$
$$(x, y, z) \longmapsto (x, y, 0).$$

We have seen that for this linear transformation:

- the image of t is the (x, y)-plane, so $\dim \mathrm{Im}(t) = 2$;
- the kernel of t is the z-axis, so $\dim \mathrm{Ker}(t) = 1$.

Thus

$$\dim \mathrm{Im}(t) + \dim \mathrm{Ker}(t) = 2 + 1 = 3,$$

which is the dimension of the domain $\mathbb{R}^3$.

Now let t be the linear transformation

$$t : \mathbb{R}^3 \longrightarrow \mathbb{R}^2$$
$$(x, y, z) \longmapsto (x + 2y + 3z, 4x + y - 2z).$$

We have seen that for this linear transformation:

- the image of t is the whole of $\mathbb{R}^2$, so $\dim \mathrm{Im}(t) = 2$;
- the kernel of t is the line through $(0, 0, 0)$ and $(1, -2, 1)$, so $\dim \mathrm{Ker}(t) = 1$.

Thus

$$\dim \mathrm{Im}(t) + \dim \mathrm{Ker}(t) = 2 + 1 = 3,$$

which is the dimension of the domain $\mathbb{R}^3$.

Exercise 4.9   For each of the linear transformations t in Exercise 4.3, calculate

$$\dim \operatorname{Im}(t) + \dim \operatorname{Ker}(t)$$

and compare your answer with the dimension of the domain of t.

Use the solutions to Exercises 4.3, 4.6 and 4.7.

For each of the linear transformations in Exercise 4.9, the dimension of the image plus the dimension of the kernel is equal to the dimension of the domain. This relationship holds for all linear transformations.

Theorem 4.4 Dimension Theorem

Let $t : V \longrightarrow W$ be a linear transformation. Then

$$\dim \operatorname{Im}(t) + \dim \operatorname{Ker}(t) = \dim V.$$

The proof is given at the end of the section.

The Dimension Theorem is an important result and has several applications. For example, using the Dimension Theorem we can obtain information on whether a linear transformation $t : V \longrightarrow W$ is one-one and/or onto.

Properties 4.1 and 4.2 state that:

t is onto if and only if $\operatorname{Im}(t) = W$;

t is one-one if and only if $\operatorname{Ker}(t) = \{\mathbf{0}\}$.

Suppose that $t : V \longrightarrow W$ is a linear transformation from the n-dimensional vector space V to the m-dimensional vector space W. We consider three cases: $n > m$, $n < m$ and $n = m$.

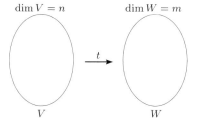

Case (a): $n > m$

Since the image of t is a subspace of W, we have $\dim \operatorname{Im}(t) \leq m$. It follows from the Dimension Theorem that

$$\dim \operatorname{Ker}(t) = \dim V - \dim \operatorname{Im}(t) \geq n - m > 0.$$

Thus $\operatorname{Ker}(t) \neq \{\mathbf{0}\}$, so t is not one-one.

For example, the linear transformation

$$t : \mathbb{R}^3 \longrightarrow \mathbb{R}^2$$
$$(x, y, z) \longmapsto (2x + y, x + z)$$

is not one-one, since the dimension of the codomain (which is 2) is less than the dimension of the domain (which is 3).

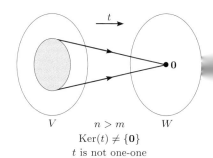

Case (b): $n < m$

By the Dimension Theorem,

$$\dim \operatorname{Im}(t) = \dim V - \dim \operatorname{Ker}(t) \leq n < m.$$

Thus $\operatorname{Im}(t)$ is not the whole of the m-dimensional vector space W, so t is not onto.

For example, the linear transformation

$$t : \mathbb{R}^2 \longrightarrow \mathbb{R}^3$$
$$(x, y) \longmapsto (2x, x + y, y)$$

is not onto, since the dimension of the codomain (which is 3) is greater than the dimension of the domain (which is 2).

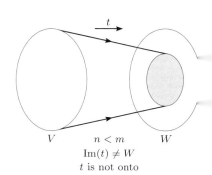

Case (c): $n = m$

By the Dimension Theorem,

$$\dim \operatorname{Im}(t) + \dim \operatorname{Ker}(t) = \dim V = n = m.$$

There are two possibilities:

EITHER $\dim \operatorname{Ker}(t) = 0$ and $\dim \operatorname{Im}(t) = n = m$,

OR $\dim \operatorname{Ker}(t) > 0$ and $\dim \operatorname{Im}(t) < m$.

If $\dim \operatorname{Ker}(t) = 0$ and $\dim \operatorname{Im}(t) = n = m$, then

$$\operatorname{Ker}(t) = \{\mathbf{0}\} \quad \text{and} \quad \operatorname{Im}(t) = W.$$

Thus t is both one-one and onto.

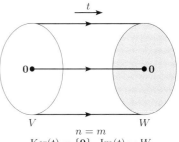

$n = m$
$\operatorname{Ker}(t) = \{\mathbf{0}\}$, $\operatorname{Im}(t) = W$
t is both one-one and onto

For example, consider the linear transformation from Exercise 4.3(a),

$$t : \mathbb{R}^2 \longrightarrow \mathbb{R}^2$$
$$(x, y) \longmapsto (x, 2x + y).$$

Here the domain and codomain both have dimension 2, and $\dim \operatorname{Im}(t) = 2$. This is equal to the dimension of the codomain, so t is both one-one and onto.

If, on the other hand, $\dim \operatorname{Ker}(t) > 0$ and $\dim \operatorname{Im}(t) < m$, then

$$\operatorname{Ker}(t) \neq \{\mathbf{0}\} \quad \text{and} \quad \operatorname{Im}(t) \text{ is not the whole of } W.$$

Thus t is neither one-one nor onto.

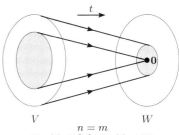

$n = m$
$\operatorname{Ker}(t) \neq \{\mathbf{0}\}$, $\operatorname{Im}(t) \neq W$
t is neither one-one nor onto

For example, consider the linear transformation from Exercise 4.3(b),

$$t : \mathbb{R}^3 \longrightarrow \mathbb{R}^3$$
$$(x, y, z) \longmapsto (x + 2y + 3z, x + z, x + y + 2z).$$

Here the domain and codomain both have dimension 3, and $\dim \operatorname{Im}(t) = 2$. This is less than the dimension of the codomain of t, thus t is neither one-one nor onto.

We summarise these findings below.

Theorem 4.5   Let $t : V \longrightarrow W$ be a linear transformation from an n-dimensional vector space V to an m-dimensional vector space W.

(a) If $n > m$, then t is not one-one ($\operatorname{Ker}(t) \neq \{\mathbf{0}\}$).

(b) If $n < m$, then t is not onto ($\operatorname{Im}(t) \neq W$).

(c) If $n = m$, then

EITHER t is both one-one and onto:
$\operatorname{Ker}(t) = \{\mathbf{0}\}$ and $\operatorname{Im}(t) = W$;

OR t is neither one-one nor onto:
$\operatorname{Ker}(t) \neq \{\mathbf{0}\}$ and $\operatorname{Im}(t) \neq W$.

Exercise 4.10   What can we deduce from Theorem 4.5 about the following linear transformations?

(a) $t : \mathbb{R}^2 \longrightarrow \mathbb{R}^3$

$(x, y) \longmapsto (x, y, x + y)$

(b) $t : \mathbb{R}^2 \longrightarrow \mathbb{R}^2$

$(x, y) \longmapsto (3x, 4x + y)$

(c) $t : P_3 \longrightarrow P_2$

$p(x) \longmapsto p'(x)$

Systems of simultaneous linear equations

We now show how we can use linear transformations to obtain information on the number of solutions to a system of simultaneous linear equations.

Suppose that we want to know how many solutions there are to the following system of three simultaneous linear equations in three unknowns:

$$\begin{cases} 2x + 3y + 4z = 7, \\ x + 5y + 6z = 4, \\ 3x + 2y + 5z = 1. \end{cases}$$

This system can be written in matrix form as

$$\begin{pmatrix} 2 & 3 & 4 \\ 1 & 5 & 6 \\ 3 & 2 & 5 \end{pmatrix} \begin{pmatrix} x \\ y \\ z \end{pmatrix} = \begin{pmatrix} 7 \\ 4 \\ 1 \end{pmatrix}.$$

Now let t be the linear transformation with the matrix representation

$$t : \mathbb{R}^3 \longrightarrow \mathbb{R}^3$$

$$\begin{pmatrix} x \\ y \\ z \end{pmatrix} \longmapsto \begin{pmatrix} 2 & 3 & 4 \\ 1 & 5 & 6 \\ 3 & 2 & 5 \end{pmatrix} \begin{pmatrix} x \\ y \\ z \end{pmatrix}.$$

We see that (x, y, z) is a solution to the system of equations precisely when $t(x, y, z) = (7, 4, 1)$.

Thus the number of solutions to the system of equations is the same as the number of vectors in $\mathbb{R}^3$ that map to the vector $(7, 4, 1)$ under t.

In general, suppose that we want to know how many solutions there are to the system of m simultaneous linear equations in n unknowns with the matrix equation

$$\mathbf{Ax} = \mathbf{b}.$$

Let t be the linear transformation with the matrix representation

$$t : \mathbb{R}^n \longrightarrow \mathbb{R}^m$$

$$\mathbf{x} \longmapsto \mathbf{Ax}.$$

Then the number of solutions to the system of equations is the same as the number of vectors that map to $\mathbf{b}$ under t.

Let $\mathbf{b} \in \mathrm{Im}(t)$. Then there is some vector $\mathbf{a}$ such that $t(\mathbf{a}) = \mathbf{b}$. The solution set to the system of equations is

$$\{\mathbf{x} : \mathbf{x} = \mathbf{a} + \mathbf{k} \text{ for some } \mathbf{k} \in \mathrm{Ker}(t)\}.$$

It follows that there are three possibilities.
- If $\mathbf{b} \in \mathrm{Im}(t)$ and $\mathrm{Ker}(t) = \{\mathbf{0}\}$, then there is exactly *one* solution.
- If $\mathbf{b} \in \mathrm{Im}(t)$ and $\mathrm{Ker}(t) \neq \{\mathbf{0}\}$, then there are *infinitely many* solutions.
- If $\mathbf{b} \notin \mathrm{Im}(t)$, then there are *no* solutions.

Thus a system of simultaneous linear equations has no solutions, or one solution, or infinitely many solutions.

> $\mathrm{Ker}(t)$ is a subspace and hence contains either just the zero vector or infinitely many vectors.

> This result was stated without proof in Unit LA2, Section 1.

Exercise 4.11 How many solutions are there to the following system of three simultaneous linear equations in three unknowns?

$$\begin{cases} x + 2y + 3z = 1 \\ x \quad\quad + \ z = 1 \\ x + \ y + 2z = 1 \end{cases}$$

> Use the solutions to Exercises 4.3(b) and 4.6(b).

By considering the linear transformation

$$t : \mathbb{R}^n \longrightarrow \mathbb{R}^m$$
$$\mathbf{x} \longmapsto \mathbf{Ax},$$

we can show that the number of solutions to the system $\mathbf{Ax} = \mathbf{b}$ of m linear equations in n unknowns depends on the values of m and n. We consider three cases: $n > m$, $n < m$ and $n = m$.

Case (a): $n > m$

It follows from Theorem 4.5 that $\mathrm{Ker}(t) \neq \{\mathbf{0}\}$. Thus the equation $\mathbf{Ax} = \mathbf{b}$ has either no solutions (if $\mathbf{b} \notin \mathrm{Im}(t)$) or infinitely many solutions (if $\mathbf{b} \in \mathrm{Im}(t)$). For example, the system

$$\begin{cases} 2x + \ y + \ z = a, \\ 4x + 2y + 2z = b, \end{cases}$$

of two simultaneous equations in three unknowns has either no solutions or infinitely many solutions, depending on the values of a and b.

> For example, the system has no solutions when $a = 3$ and $b = 4$, and infinitely many solutions when $a = 2$ and $b = 4$.

Case (b): $n < m$

It follows from Theorem 4.5 that $\mathrm{Im}(t) \neq \mathbb{R}^m$. Thus there is some $\mathbf{b}$ for which the equation $\mathbf{Ax} = \mathbf{b}$ has no solutions. For example, there are some values of a, b and c for which the system

$$\begin{cases} 2x + \ y = a, \\ x + 3y = b, \\ 4x + \ y = c, \end{cases}$$

of three equations in two unknowns has no solutions.

> For example, the system has no solutions when $a = 3$, $b = 4$ and $c = 2$.

Case (c): $n = m$

It follows from Theorem 4.5 that there are two possibilities.

If $\mathrm{Ker}(t) = \{\mathbf{0}\}$ and $\mathrm{Im}(t) = \mathbb{R}^m$, then the equation $\mathbf{Ax} = \mathbf{b}$ has exactly one solution for each $\mathbf{b}$. For example, the system

$$\begin{cases} x + y = a, \\ y = b, \end{cases}$$

of two simultaneous equations in two unknowns has exactly one solution, namely $(x, y) = (a - b, b)$, for each pair of values (a, b).

If $\mathrm{Ker}(t) \neq \{\mathbf{0}\}$ and $\mathrm{Im}(t) \neq \mathbb{R}^m$, then there exist vectors $\mathbf{b}$ for which the equation $\mathbf{Ax} = \mathbf{b}$ has no solutions, and for all other $\mathbf{b}$, the equation $\mathbf{Ax} = \mathbf{b}$ has infinitely many solutions. Consider the system

$$\begin{cases} x + 2y = a, \\ 2x + 4y = b, \end{cases}$$

of two simultaneous equations in two unknowns. Since $2x + 4y = 2(x + 2y)$, these equations have no solution when $b \neq 2a$. When $b = 2a$, however, putting $y = k$ gives $(x, y) = (a - 2k, k)$, where $k \in \mathbb{R}$, as a solution to the equations; thus there are infinitely many solutions.

We summarise these results below.

Theorem 4.6   Let $\mathbf{Ax} = \mathbf{b}$ be a system of m simultaneous linear equations in n unknowns.

(a) If $n > m$, then $\mathbf{Ax} = \mathbf{b}$ has either no solutions or infinitely many solutions.

(b) If $n < m$, then there is some $\mathbf{b}$ for which $\mathbf{Ax} = \mathbf{b}$ has no solutions.

(c) If $n = m$, then:

EITHER $\mathbf{Ax} = \mathbf{b}$ has exactly one solution for each $\mathbf{b}$;

OR   there are some $\mathbf{b}$ for which $\mathbf{Ax} = \mathbf{b}$ has no solutions; for all other $\mathbf{b}$, $\mathbf{Ax} = \mathbf{b}$ has infinitely many solutions.

Exercise 4.12   What can you deduce from Theorem 4.6 about the number of solutions to each of the following systems of simultaneous linear equations?

(a) $\begin{cases} 3x + y + z = 1 \\ 4x + 2y + 4z = 3 \end{cases}$ (b) $\begin{cases} 3x + y + z = a \\ 4x + 2y + 4z = b \\ 5x + y + 6z = c \end{cases}$

Proof of the Dimension Theorem

Having seen how useful the Dimension Theorem can be, we end this section by proving it.

Theorem 4.4   Dimension Theorem

Let $t : V \longrightarrow W$ be a linear transformation. Then

$$\dim \mathrm{Im}(t) + \dim \mathrm{Ker}(t) = \dim V.$$

Proof   Let $\dim V = n$ and $\dim \mathrm{Ker}(t) = k$. We must show that

$$\dim \mathrm{Im}(t) = n - k.$$

Let $\{\mathbf{e}_1, \ldots, \mathbf{e}_k\}$ be a basis for $\mathrm{Ker}(t)$. We can extend this to give a basis $\{\mathbf{e}_1, \ldots, \mathbf{e}_n\}$ for V. We prove that

$$F = \{t(\mathbf{e}_{k+1}), \ldots, t(\mathbf{e}_n)\}$$

is a basis for $\mathrm{Im}(t)$, which shows that $\dim \mathrm{Im}(t) = n - k$.

If you are short of time, omit this proof.

We showed that this extension is possible in Unit LA3, Theorem 3.6.

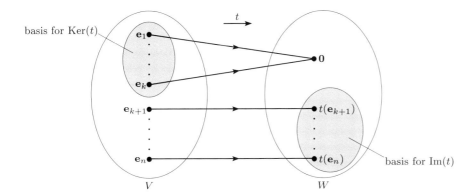

To show that F is a basis for $\mathrm{Im}(t)$, we use Strategy 3.2 in Unit LA3.

First we prove that F spans $\mathrm{Im}(t)$. We know from Subsection 4.1 that $\{t(\mathbf{e}_1), \ldots, t(\mathbf{e}_n)\}$ spans $\mathrm{Im}(t)$. Since $\mathbf{e}_1, \ldots, \mathbf{e}_k$ belong to $\mathrm{Ker}(t)$, we know that

$$t(\mathbf{e}_1) = t(\mathbf{e}_2) = \cdots = t(\mathbf{e}_k) = \mathbf{0},$$

so the span of $\{t(\mathbf{e}_1), \ldots, t(\mathbf{e}_n)\}$ is equal to the span of $\{t(\mathbf{e}_{k+1}), \ldots, t(\mathbf{e}_n)\}$. Thus F spans $\mathrm{Im}(t)$.

Next we show that F is a linearly independent set. We must show that if

$$\alpha_{k+1}t(\mathbf{e}_{k+1}) + \alpha_{k+2}t(\mathbf{e}_{k+2}) + \cdots + \alpha_n t(\mathbf{e}_n) = \mathbf{0},$$

then

$$\alpha_{k+1} = \alpha_{k+2} = \cdots = \alpha_n = 0.$$

Since t is a linear transformation, we have

$$\alpha_{k+1}t(\mathbf{e}_{k+1}) + \cdots + \alpha_n t(\mathbf{e}_n) = t(\alpha_{k+1}\mathbf{e}_{k+1} + \cdots + \alpha_n\mathbf{e}_n).$$

So if $\alpha_{k+1}t(\mathbf{e}_{k+1}) + \cdots + \alpha_n t(\mathbf{e}_n) = \mathbf{0}$, then

$$t(\alpha_{k+1}\mathbf{e}_{k+1} + \cdots + \alpha_n\mathbf{e}_n) = \mathbf{0},$$

thus

$$\alpha_{k+1}\mathbf{e}_{k+1} + \cdots + \alpha_n\mathbf{e}_n \in \mathrm{Ker}(t).$$

Since $\{\mathbf{e}_1, \ldots, \mathbf{e}_k\}$ is a basis for $\mathrm{Ker}(t)$, there exist real numbers $\alpha_1, \ldots, \alpha_k$ such that

$$\alpha_{k+1}\mathbf{e}_{k+1} + \cdots + \alpha_n\mathbf{e}_n = \alpha_1\mathbf{e}_1 + \cdots + \alpha_k\mathbf{e}_k,$$

so

$$\alpha_1\mathbf{e}_1 + \cdots + \alpha_k\mathbf{e}_k - \alpha_{k+1}\mathbf{e}_{k+1} - \cdots - \alpha_n\mathbf{e}_n = \mathbf{0}.$$

Since $\{\mathbf{e}_1, \ldots, \mathbf{e}_n\}$ is a basis for V, it follows that

$$\alpha_1 = \cdots = \alpha_k = -\alpha_{k+1} = \cdots = -\alpha_n = 0.$$

Thus

$$\alpha_{k+1} = \alpha_{k+2} = \cdots = \alpha_n = 0,$$

as required.

Thus F is a basis for $\mathrm{Im}(t)$, so $\dim \mathrm{Im}(t) + \dim \mathrm{Ker}(t) = \dim V$. ∎

Further exercises

Exercise 4.13   Let t be the linear transformation

$$t : \mathbb{R}^3 \longrightarrow \mathbb{R}^2$$
$$(x, y, z) \longmapsto (x + y, x - z).$$

(a) What can you deduce from the Dimension Theorem about whether t is one-one and/or onto?

(b) Find the kernel of t and the dimension of the kernel.

(c) Hence find $\operatorname{Im}(t)$.

(d) Is t one-one and/or onto?

Exercise 4.14   Let t be the linear transformation

$$t : \mathbb{R}^3 \longrightarrow \mathbb{R}^3$$
$$(x, y, z) \longmapsto (x + 2y + 3z, y + z, x + z).$$

(a) Find $\operatorname{Im}(t)$.

(b) Find $\operatorname{Ker}(t)$.

(c) Use your answers to parts (a) and (b) to determine the number of solutions to the following system of linear equations.

$$\begin{cases} x + 2y + 3z = 4 \\ \quad\;\; y + \;\; z = 1 \\ x \quad\;\; + \;\; z = 2 \end{cases}$$

Exercise 4.15   Let $V = \{f(x) : f(x) = ae^x \cos x + be^x \sin x, \; a, b \in \mathbb{R}\}$, and let t be the linear transformation

$$t : V \longrightarrow V$$
$$f(x) \longmapsto f'(x).$$

(a) Find a basis for $\operatorname{Im}(t)$, and state the dimension of $\operatorname{Im}(t)$.

(b) Hence find $\operatorname{Ker}(t)$.

Solutions to the exercises

1.1 (a)

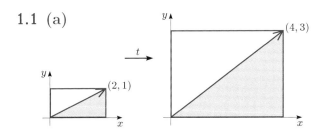

This is a $(2, 3)$-stretching.

(b)

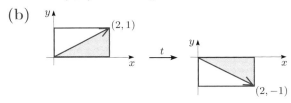

This is a reflection in the x-axis; it is also a $(1, -1)$-stretching.

(c)

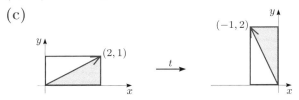

This is a rotation through an angle $\pi/2$.

1.2 We use Strategy 1.1.

(a) First $t(\mathbf{0}) = \mathbf{0}$, so t may be a linear transformation.

Next we check whether t satisfies LT1:
$$t(\mathbf{v}_1 + \mathbf{v}_2) = t(\mathbf{v}_1) + t(\mathbf{v}_2), \quad \text{for all } \mathbf{v}_1, \mathbf{v}_2 \in \mathbb{R}^2.$$
In $\mathbb{R}^2$, let $\mathbf{v}_1 = (x_1, y_1)$ and $\mathbf{v}_2 = (x_2, y_2)$. Then
$$
\begin{aligned}
t(\mathbf{v}_1 + \mathbf{v}_2) &= t(x_1 + x_2, y_1 + y_2) \\
&= (x_1 + x_2 + 3(y_1 + y_2), y_1 + y_2) \\
&= (x_1 + x_2 + 3y_1 + 3y_2, y_1 + y_2)
\end{aligned}
$$
and
$$
\begin{aligned}
t(\mathbf{v}_1) + t(\mathbf{v}_2) &= (x_1 + 3y_1, y_1) + (x_2 + 3y_2, y_2) \\
&= (x_1 + x_2 + 3y_1 + 3y_2, y_1 + y_2).
\end{aligned}
$$
These expressions are equal, so LT1 is satisfied.

Finally, we check whether t satisfies LT2:
$$t(\alpha\mathbf{v}) = \alpha\, t(\mathbf{v}), \quad \text{for all } \mathbf{v} \in \mathbb{R}^2,\ \alpha \in \mathbb{R}.$$
Let $\mathbf{v} = (x, y)$ be a vector in $\mathbb{R}^2$ and let $\alpha \in \mathbb{R}$. Then
$$t(\alpha\mathbf{v}) = t(\alpha x, \alpha y) = (\alpha x + 3\alpha y, \alpha y)$$
and
$$\alpha\, t(\mathbf{v}) = \alpha(x + 3y, y) = (\alpha x + 3\alpha y, \alpha y).$$
These expressions are equal, so LT2 is satisfied.

Since both LT1 and LT2 are satisfied, t is a linear transformation.

(b) Since $t(\mathbf{0}) = t(0, 0) = (2, 1) \neq \mathbf{0}$, it follows from Strategy 1.1 that t is not a linear transformation.

1.3 We use Strategy 1.1.

(a) First $t(\mathbf{0}) = \mathbf{0}$, so t may be a linear transformation.

Next we check whether t satisfies LT1:
$$t(\mathbf{v}_1 + \mathbf{v}_2) = t(\mathbf{v}_1) + t(\mathbf{v}_2), \quad \text{for all } \mathbf{v}_1, \mathbf{v}_2 \in \mathbb{R}^2.$$
In $\mathbb{R}^2$, let $\mathbf{v}_1 = (x_1, y_1)$ and $\mathbf{v}_2 = (x_2, y_2)$. Then
$$
\begin{aligned}
&t(\mathbf{v}_1 + \mathbf{v}_2) \\
&= t(x_1 + x_2, y_1 + y_2) \\
&= (x_1 + x_2, y_1 + y_2, x_1 + x_2, y_1 + y_2)
\end{aligned}
$$
and
$$
\begin{aligned}
&t(\mathbf{v}_1) + t(\mathbf{v}_2) \\
&= (x_1, y_1, x_1, y_1) + (x_2, y_2, x_2, y_2) \\
&= (x_1 + x_2, y_1 + y_2, x_1 + x_2, y_1 + y_2).
\end{aligned}
$$
These expressions are equal, so LT1 is satisfied.

Finally, we check whether t satisfies LT2:
$$t(\alpha\mathbf{v}) = \alpha\, t(\mathbf{v}), \quad \text{for all } \mathbf{v} \in \mathbb{R}^2,\ \alpha \in \mathbb{R}.$$
Let $\mathbf{v} = (x, y)$ be a vector in $\mathbb{R}^2$ and let $\alpha \in \mathbb{R}$. Then
$$t(\alpha\mathbf{v}) = t(\alpha x, \alpha y) = (\alpha x, \alpha y, \alpha x, \alpha y)$$
and
$$\alpha\, t(\mathbf{v}) = \alpha(x, y, x, y) = (\alpha x, \alpha y, \alpha x, \alpha y).$$
These expressions are equal, so LT2 is satisfied.

Since both LT1 and LT2 are satisfied, t is a linear transformation.

(b) First $t(\mathbf{0}) = \mathbf{0}$, so t may be a linear transformation.

Next we check whether t satisfies LT1:
$$t(\mathbf{v}_1 + \mathbf{v}_2) = t(\mathbf{v}_1) + t(\mathbf{v}_2), \quad \text{for all } \mathbf{v}_1, \mathbf{v}_2 \in \mathbb{R}^3.$$
In $\mathbb{R}^3$, let $\mathbf{v}_1 = (x_1, y_1, z_1)$ and $\mathbf{v}_2 = (x_2, y_2, z_2)$. Then
$$
\begin{aligned}
t(\mathbf{v}_1 + \mathbf{v}_2) &= t(x_1 + x_2, y_1 + y_2, z_1 + z_2) \\
&= (x_1 + x_2)^2 \\
&= x_1^2 + x_2^2 + 2x_1 x_2
\end{aligned}
$$
and
$$t(\mathbf{v}_1) + t(\mathbf{v}_2) = x_1^2 + x_2^2.$$
These expressions are not equal in general, so LT1 is not satisfied.

Thus t is not a linear transformation.

(c) Since $t(\mathbf{0}) = t(0, 0, 0) = (0, 0, 0, 1) \neq \mathbf{0}$, it follows that t is not a linear transformation.

1.4 First we show that t satisfies LT1:
$$t(\mathbf{v}_1 + \mathbf{v}_2) = t(\mathbf{v}_1) + t(\mathbf{v}_2), \quad \text{for all } \mathbf{v}_1, \mathbf{v}_2 \in \mathbb{R}^3.$$
In $\mathbb{R}^3$, let $\mathbf{v}_1 = (x_1, y_1, z_1)$ and $\mathbf{v}_2 = (x_2, y_2, z_2)$. Then
$$
\begin{aligned}
&t(\mathbf{v}_1 + \mathbf{v}_2) \\
&= t(x_1 + x_2, y_1 + y_2, z_1 + z_2) \\
&= ((x_1 + x_2)\cos\theta - (y_1 + y_2)\sin\theta, \\
&\quad (x_1 + x_2)\sin\theta + (y_1 + y_2)\cos\theta, z_1 + z_2)
\end{aligned}
$$
and
$$
\begin{aligned}
&t(\mathbf{v}_1) + t(\mathbf{v}_2) \\
&= (x_1\cos\theta - y_1\sin\theta, x_1\sin\theta + y_1\cos\theta, z_1) \\
&\quad + (x_2\cos\theta - y_2\sin\theta, x_2\sin\theta + y_2\cos\theta, z_2) \\
&= ((x_1 + x_2)\cos\theta - (y_1 + y_2)\sin\theta, \\
&\quad (x_1 + x_2)\sin\theta + (y_1 + y_2)\cos\theta, z_1 + z_2).
\end{aligned}
$$
These expressions are equal, so LT1 is satisfied.
Next we show that t satisfies LT2:
$$t(\alpha\mathbf{v}) = \alpha\, t(\mathbf{v}), \quad \text{for all } \mathbf{v} \in \mathbb{R}^3, \ \alpha \in \mathbb{R}.$$
Let $\mathbf{v} = (x, y, z)$ be a vector in $\mathbb{R}^3$ and let $\alpha \in \mathbb{R}$. Then
$$
\begin{aligned}
t(\alpha\mathbf{v}) &= t(\alpha x, \alpha y, \alpha z) \\
&= (\alpha x\cos\theta - \alpha y\sin\theta, \alpha x\sin\theta + \alpha y\cos\theta, \alpha z)
\end{aligned}
$$
and
$$
\begin{aligned}
\alpha\, t(\mathbf{v}) &= \alpha(x\cos\theta - y\sin\theta, x\sin\theta + y\cos\theta, z) \\
&= (\alpha x\cos\theta - \alpha y\sin\theta, \alpha x\sin\theta + \alpha y\cos\theta, \alpha z).
\end{aligned}
$$
These expressions are equal, so LT2 is satisfied.
Since LT1 and LT2 are satisfied, t is a linear transformation.

1.5 We use Strategy 1.1.
Since the zero element of P_3 is $p(x) = 0$, we have $t(\mathbf{0}) = \mathbf{0}$, so t may be a linear transformation.
Next we check whether t satisfies LT1:
$$
\begin{aligned}
&t(p(x) + q(x)) = t(p(x)) + t(q(x)), \\
&\quad \text{for all } p(x), q(x) \in P_3.
\end{aligned}
$$
Let $p(x), q(x) \in P_3$. Then
$$t(p(x) + q(x)) = p(x) + q(x) + p(2) + q(2)$$
and
$$
\begin{aligned}
t(p(x)) + t(q(x)) &= p(x) + p(2) + q(x) + q(2) \\
&= p(x) + q(x) + p(2) + q(2).
\end{aligned}
$$
These expressions are equal, so LT1 is satisfied.
Finally, we check whether t satisfies LT2:
$$t(\alpha p(x)) = \alpha\, t(p(x)), \quad \text{for all } p(x) \in P_3, \ \alpha \in \mathbb{R}.$$
Let $p(x) \in P_3$ and $\alpha \in \mathbb{R}$. Then
$$t(\alpha p(x)) = \alpha p(x) + \alpha p(2)$$
and
$$\alpha\, t(p(x)) = \alpha(p(x) + p(2)) = \alpha p(x) + \alpha p(2).$$
These expressions are equal, so LT2 is satisfied.
Since both LT1 and LT2 are satisfied, t is a linear transformation.

1.6 First we show that i_V satisfies LT1:
$$i_V(\mathbf{v}_1 + \mathbf{v}_2) = i_V(\mathbf{v}_1) + i_V(\mathbf{v}_2), \quad \text{for all } \mathbf{v}_1, \mathbf{v}_2 \in V.$$
Let $\mathbf{v}_1, \mathbf{v}_2 \in V$. Then
$$i_V(\mathbf{v}_1 + \mathbf{v}_2) = \mathbf{v}_1 + \mathbf{v}_2$$
and
$$i_V(\mathbf{v}_1) + i_V(\mathbf{v}_2) = \mathbf{v}_1 + \mathbf{v}_2.$$
These expressions are equal, so LT1 is satisfied.
Next we show that i_V satisfies LT2:
$$i_V(\alpha\mathbf{v}) = \alpha\, i_V(\mathbf{v}), \quad \text{for all } \mathbf{v} \in V, \ \alpha \in \mathbb{R}.$$
Let $\mathbf{v} \in V$ and $\alpha \in \mathbb{R}$. Then
$$i_V(\alpha\mathbf{v}) = \alpha\mathbf{v}$$
and
$$\alpha\, i_V(\mathbf{v}) = \alpha\mathbf{v}.$$
These expressions are equal, so LT2 is satisfied.
Since LT1 and LT2 are satisfied, t is a linear transformation.

1.7 We can write any vector (x, y) in $\mathbb{R}^2$ in the form
$$(x, y) = x(1, 0) + y(0, 1).$$
It follows from Theorem 1.3 that
$$
\begin{aligned}
q_\phi(x, y) &= x(\cos 2\phi, \sin 2\phi) + y(\sin 2\phi, -\cos 2\phi) \\
&= (x\cos 2\phi + y\sin 2\phi, x\sin 2\phi - y\cos 2\phi).
\end{aligned}
$$

1.8 We use Strategy 1.1.
(a) First $t(\mathbf{0}) = \mathbf{0}$, so t may be a linear transformation.
Next we check whether t satisfies LT1:
$$t(\mathbf{v}_1 + \mathbf{v}_2) = t(\mathbf{v}_1) + t(\mathbf{v}_2), \quad \text{for all } \mathbf{v}_1, \mathbf{v}_2 \in \mathbb{R}^2.$$
In $\mathbb{R}^2$, let $\mathbf{v}_1 = (x_1, y_1)$ and $\mathbf{v}_2 = (x_2, y_2)$. Then
$$
\begin{aligned}
&t(\mathbf{v}_1 + \mathbf{v}_2) \\
&= t(x_1 + x_2, y_1 + y_2) \\
&= (3(x_1 + x_2) + y_1 + y_2, 2(x_1 + x_2) - (y_1 + y_2))
\end{aligned}
$$
and
$$
\begin{aligned}
&t(\mathbf{v}_1) + t(\mathbf{v}_2) \\
&= (3x_1 + y_1, 2x_1 - y_1) + (3x_2 + y_2, 2x_2 - y_2) \\
&= (3(x_1 + x_2) + y_1 + y_2, 2(x_1 + x_2) - (y_1 + y_2)).
\end{aligned}
$$
These expressions are equal, so LT1 is satisfied.
Finally, we check whether t satisfies LT2:
$$t(\alpha\mathbf{v}) = \alpha\, t(\mathbf{v}), \quad \text{for all } \mathbf{v} \in \mathbb{R}^2, \ \alpha \in \mathbb{R}.$$
Let $\mathbf{v} = (x, y)$ be a vector in $\mathbb{R}^2$, and let $\alpha \in \mathbb{R}$. Then
$$
\begin{aligned}
t(\alpha\mathbf{v}) &= t(\alpha x, \alpha y) \\
&= (3\alpha x + \alpha y, 2\alpha x - \alpha y)
\end{aligned}
$$
and
$$
\begin{aligned}
\alpha\, t(\mathbf{v}) &= \alpha(3x + y, 2x - y) \\
&= (3\alpha x + \alpha y, 2\alpha x - \alpha y).
\end{aligned}
$$
These expressions are equal, so LT2 is satisfied.
Since LT1 and LT2 are satisfied, t is a linear transformation.

(b) Since $t(\mathbf{0}) = t(0,0) = (1,0) \neq \mathbf{0}$, it follows that t is not a linear transformation.

1.9 We use Strategy 1.1.

(a) First $t(\mathbf{0}) = \mathbf{0}$, so t may be a linear transformation.

Next we check whether t satisfies LT1:
$$t(\mathbf{v}_1 + \mathbf{v}_2) = t(\mathbf{v}_1) + t(\mathbf{v}_2), \quad \text{for all } \mathbf{v}_1, \mathbf{v}_2 \in \mathbb{R}^2.$$
In $\mathbb{R}^2$, let $\mathbf{v}_1 = (x_1, y_1)$ and $\mathbf{v}_2 = (x_2, y_2)$. Then
$$\begin{aligned} t(\mathbf{v}_1 + \mathbf{v}_2) &= t(x_1 + x_2, y_1 + y_2) \\ &= ((x_1 + x_2)^2, y_1 + y_2, y_1 + y_2) \end{aligned}$$
and
$$\begin{aligned} t(\mathbf{v}_1) + t(\mathbf{v}_2) &= (x_1^2, y_1, y_1) + (x_2^2, y_2, y_2) \\ &= (x_1^2 + x_2^2, y_1 + y_2, y_1 + y_2). \end{aligned}$$
These expressions are not equal in general, so LT1 is not satisfied.

Thus t is not a linear transformation.

(b) First $t(\mathbf{0}) = \mathbf{0}$, so t may be a linear transformation.

Next we check whether t satisfies LT1:
$$t(\mathbf{v}_1 + \mathbf{v}_2) = t(\mathbf{v}_1) + t(\mathbf{v}_2), \quad \text{for all } \mathbf{v}_1, \mathbf{v}_2 \in \mathbb{R}^3.$$
In $\mathbb{R}^3$, let $\mathbf{v}_1 = (x_1, y_1, z_1)$ and $\mathbf{v}_2 = (x_2, y_2, z_2)$. Then
$$\begin{aligned} &t(\mathbf{v}_1 + \mathbf{v}_2) \\ &= t(x_1 + x_2, y_1 + y_2, z_1 + z_2) \\ &= (x_1 + x_2 + y_1 + y_2, x_1 + x_2 - z_1 - z_2) \end{aligned}$$
and
$$\begin{aligned} &t(\mathbf{v}_1) + t(\mathbf{v}_2) \\ &= (x_1 + y_1, x_1 - z_1) + (x_2 + y_2, x_2 - z_2) \\ &= (x_1 + x_2 + y_1 + y_2, x_1 + x_2 - z_1 - z_2). \end{aligned}$$
These expressions are equal, so LT1 is satisfied.

Finally, we check whether t satisfies LT2:
$$t(\alpha \mathbf{v}) = \alpha\, t(\mathbf{v}), \quad \text{for all } \mathbf{v} \in \mathbb{R}^3, \ \alpha \in \mathbb{R}.$$
Let $\mathbf{v} = (x, y, z)$ be a vector in $\mathbb{R}^3$, and let $\alpha \in \mathbb{R}$. Then
$$\begin{aligned} t(\alpha \mathbf{v}) &= t(\alpha x, \alpha y, \alpha z) \\ &= (\alpha x + \alpha y, \alpha x - \alpha z) \end{aligned}$$
and
$$\begin{aligned} \alpha\, t(\mathbf{v}) &= \alpha(x + y, x - z) \\ &= (\alpha x + \alpha y, \alpha x - \alpha z). \end{aligned}$$
These expressions are equal, so LT2 is satisfied.

Since LT1 and LT2 are satisfied, t is a linear transformation.

(c) Since $t(\mathbf{0}) = t(0,0) = (0,0,1,0) \neq \mathbf{0}$, it follows that t is not a linear transformation.

1.10 We use Strategy 1.1.

(a) First $t(\mathbf{0}) = \mathbf{0}$, so t may be a linear transformation.

Next we check whether t satisfies LT1:
$$t(p(x) + q(x)) = t(p(x)) + t(q(x)),$$
$$\text{for all } p(x), q(x) \in P_3.$$
Let $p(x), q(x) \in P_3$. Then
$$t(p(x) + q(x)) = p(x) + q(x) + p'(x) + q'(x)$$
and
$$\begin{aligned} t(p(x)) + t(q(x)) &= p(x) + p'(x) + q(x) + q'(x) \\ &= p(x) + q(x) + p'(x) + q'(x). \end{aligned}$$
These expressions are equal, so LT1 is satisfied.

Finally, we check whether t satisfies LT2:
$$t(\alpha p(x)) = \alpha\, t(p(x)), \quad \text{for all } p(x) \in P_3, \ \alpha \in \mathbb{R}.$$
Let $p(x) \in P_3$ and $\alpha \in \mathbb{R}$. Then
$$t(\alpha p(x)) = \alpha p(x) + \alpha p'(x)$$
and
$$\begin{aligned} \alpha\, t(p(x)) &= \alpha(p(x) + p'(x)) \\ &= \alpha p(x) + \alpha p'(x). \end{aligned}$$
These expressions are equal, so LT2 is satisfied.

Since both LT1 and LT2 are satisfied, t is a linear transformation.

(b) The zero element of P_3 is $p(x) = 0$ which maps to the polynomial $q(x) = 2$. Thus $t(\mathbf{0}) \neq \mathbf{0}$, so it follows that t is not a linear transformation.

1.11 We use Strategy 1.1.

(a) First $t(\mathbf{0}) = \mathbf{0}$, so t may be a linear transformation.

Next we check whether t satisfies LT1:
$$t(f(x) + g(x)) = t(f(x)) + t(g(x)),$$
$$\text{for all } f(x), g(x) \in V.$$
Let $f(x), g(x) \in V$. Then
$$t(f(x) + g(x)) = f'(x) + g'(x)$$
and
$$t(f(x)) + t(g(x)) = f'(x) + g'(x).$$
These expressions are equal, so LT1 is satisfied.

Finally, we check whether t satisfies LT2:
$$t(\alpha f(x)) = \alpha\, t(f(x)), \quad \text{for all } f(x) \in V, \ \alpha \in \mathbb{R}.$$
Let $f(x) \in V$ and $\alpha \in \mathbb{R}$. Then
$$t(\alpha f(x)) = \alpha f'(x)$$
and
$$\alpha\, t(f(x)) = \alpha f'(x).$$
These expressions are equal, so LT2 is satisfied.

Since both LT1 and LT2 are satisfied, t is a linear transformation.

(b) Since $t(\mathbf{0}) = t(0e^x \cos x + 0e^x \sin x) = (0,1) \neq \mathbf{0}$, it follows that t is not a linear transformation.

2.1 (a) Here $E = \{(3,1),(2,1)\}$. Therefore,
$$\mathbf{v} = (3,1) = 1(3,1) + 0(2,1),$$
so it follows that
$$\mathbf{v}_E = (1,0)_E.$$

(b) Here $E = \{(1,2),(2,1)\}$. We must find $a, b \in \mathbb{R}$ such that
$$(3,1) = (a,b)_E.$$
Since
$$(a,b)_E = a(1,2) + b(2,1) = (a + 2b, 2a + b),$$
it follows that a and b are the solutions of
$$\begin{cases} a + 2b = 3, \\ 2a + b = 1, \end{cases}$$
that is, $a = -\frac{1}{3}$ and $b = \frac{5}{3}$, so
$$\mathbf{v}_E = \left(-\tfrac{1}{3}, \tfrac{5}{3}\right)_E.$$

2.2 (a) Here $E = \{1, x\}$. Therefore,
$$p(x) = 2 + 3x = 2 \times (1) + 3 \times (x),$$
so the E-coordinate representation of $p(x)$ is
$$(2,3)_E.$$

(b) Here $E = \{1, 4 + 6x\}$. Therefore,
$$p(x) = 2 + 3x = 0 \times (1) + \tfrac{1}{2} \times (4 + 6x),$$
so the E-coordinate representation of $p(x)$ is
$$\left(0, \tfrac{1}{2}\right)_E.$$

(c) Here $E = \{2x, 1 + 4x\}$. We must find $a, b \in \mathbb{R}$ such that
$$p(x) = 2 + 3x = (a,b)_E.$$
Since
$$(a,b)_E = a \times 2x + b \times (1 + 4x) = b + (2a + 4b)x,$$
it follows that a and b are the solutions of
$$\begin{cases} b = 2, \\ 2a + 4b = 3, \end{cases}$$
that is, $a = -\frac{5}{2}$ and $b = 2$. Thus the E-coordinate representation of $p(x)$ is
$$\left(-\tfrac{5}{2}, 2\right)_E.$$

2.3 (a) We have
$$\begin{pmatrix} 1 \\ 0 \end{pmatrix} \longmapsto \begin{pmatrix} 3 & 0 \\ 0 & 2 \end{pmatrix} \begin{pmatrix} 1 \\ 0 \end{pmatrix} = \begin{pmatrix} 3 \\ 0 \end{pmatrix},$$
so $t(1,0) = (3,0)$.
Similarly,
$$\begin{pmatrix} 0 \\ 1 \end{pmatrix} \longmapsto \begin{pmatrix} 3 & 0 \\ 0 & 2 \end{pmatrix} \begin{pmatrix} 0 \\ 1 \end{pmatrix} = \begin{pmatrix} 0 \\ 2 \end{pmatrix},$$
so $t(0,1) = (0,2)$.
Thus the coordinates of $t(1,0)$ form the first column of the matrix of t, and the coordinates of $t(0,1)$ form the second column of the matrix of t.

(b) We have
$$\begin{pmatrix} 1 \\ 0 \end{pmatrix} \longmapsto \begin{pmatrix} \frac{1}{\sqrt{2}} & -\frac{1}{\sqrt{2}} \\ \frac{1}{\sqrt{2}} & \frac{1}{\sqrt{2}} \end{pmatrix} \begin{pmatrix} 1 \\ 0 \end{pmatrix} = \begin{pmatrix} \frac{1}{\sqrt{2}} \\ \frac{1}{\sqrt{2}} \end{pmatrix},$$
so $t(1,0) = \left(\frac{1}{\sqrt{2}}, \frac{1}{\sqrt{2}}\right)$.
Similarly,
$$\begin{pmatrix} 0 \\ 1 \end{pmatrix} \longmapsto \begin{pmatrix} \frac{1}{\sqrt{2}} & -\frac{1}{\sqrt{2}} \\ \frac{1}{\sqrt{2}} & \frac{1}{\sqrt{2}} \end{pmatrix} \begin{pmatrix} 0 \\ 1 \end{pmatrix} = \begin{pmatrix} -\frac{1}{\sqrt{2}} \\ \frac{1}{\sqrt{2}} \end{pmatrix},$$
so $t(0,1) = \left(-\frac{1}{\sqrt{2}}, \frac{1}{\sqrt{2}}\right)$.
As in part (a), the coordinates of $t(1,0)$ form the first column of the matrix of t, and the coordinates of $t(0,1)$ form the second column of the matrix of t.

2.4 We follow Strategy 2.1.

(a) We find the images of the vectors in the domain basis $E = \{(1,0),(0,1)\}$:
$$t(1,0) = (1,0), \quad t(0,1) = (3,1).$$
We find the F-coordinates of each of these image vectors, where $F = \{(1,0),(0,1)\}$:
$$t(1,0) = (1,0)_F, \quad t(0,1) = (3,1)_F.$$
The matrix of t with respect to the standard bases for the domain and codomain is
$$\mathbf{A} = \begin{pmatrix} 1 & 3 \\ 0 & 1 \end{pmatrix}.$$
(Remember that the image vectors in step 2 form the columns of the matrix in step 3).

Hence the matrix representation of t with respect to the standard bases for the domain and codomain is
$$\begin{pmatrix} x \\ y \end{pmatrix} \longmapsto \begin{pmatrix} 1 & 3 \\ 0 & 1 \end{pmatrix} \begin{pmatrix} x \\ y \end{pmatrix} = \begin{pmatrix} x + 3y \\ y \end{pmatrix}.$$

(b) We find the images of the vectors in the domain basis $E = \{1, x, x^2\}$:
$$t(1) = 1 + 1 = 2, \quad t(x) = x + 2,$$
$$t(x^2) = x^2 + 2^2 = x^2 + 4.$$
We find the F-coordinates of each of these image vectors, where $F = \{1, x, x^2\}$:
$$t(1) = (2,0,0)_F, \quad t(x) = (2,1,0)_F,$$
$$t(x^2) = (4,0,1)_F.$$
The matrix of t with respect to the standard bases for the domain and codomain is
$$\mathbf{A} = \begin{pmatrix} 2 & 2 & 4 \\ 0 & 1 & 0 \\ 0 & 0 & 1 \end{pmatrix}.$$
Hence the matrix representation of t with respect to the standard bases for the domain and codomain is
$$\begin{pmatrix} a \\ b \\ c \end{pmatrix} \longmapsto \begin{pmatrix} 2 & 2 & 4 \\ 0 & 1 & 0 \\ 0 & 0 & 1 \end{pmatrix} \begin{pmatrix} a \\ b \\ c \end{pmatrix} = \begin{pmatrix} 2a + 2b + 4c \\ b \\ c \end{pmatrix}$$

2.5 We follow Strategy 2.1.

(a) We find the images of the vectors in the domain basis $E = \{(1,0,1),(1,0,0),(1,1,1)\}$:
$$t(1,0,1) = (1,0), \quad t(1,0,0) = (1,0),$$
$$t(1,1,1) = (1,1).$$
We find the F-coordinates of each of these image vectors, where $F = \{(1,0),(0,1)\}$:
$$t(1,0,1) = (1,0)_F, \quad t(1,0,0) = (1,0)_F,$$
$$t(1,1,1) = (1,1)_F.$$
The matrix of t with respect to the bases E and F is
$$\mathbf{A} = \begin{pmatrix} 1 & 1 & 1 \\ 0 & 0 & 1 \end{pmatrix}.$$

(b) We find the images of the vectors in the domain basis $E = \{(1,0,0),(0,1,0),(0,0,1)\}$:
$$t(1,0,0) = (1,0), \quad t(0,1,0) = (0,1),$$
$$t(0,0,1) = (0,0).$$
We find the F-coordinates of each of these image vectors, where $F = \{(2,1),(1,1)\}$.

We first find $a, b \in \mathbb{R}$ such that
$$t(1,0,0) = (1,0) = (a,b)_F.$$
Since
$$(a,b)_F = a(2,1) + b(1,1) = (2a+b, a+b),$$
it follows that a and b are the solutions of
$$\begin{cases} 2a + b = 1, \\ a + b = 0, \end{cases}$$
that is, $a = 1$ and $b = -1$. So
$$t(1,0,0) = (1,-1)_F.$$
We next find $c, d \in \mathbb{R}$ such that
$$t(0,1,0) = (0,1) = (c,d)_F.$$
Since
$$(c,d)_F = c(2,1) + d(1,1) = (2c+d, c+d),$$
it follows that c and d are the solutions of
$$\begin{cases} 2c + d = 0, \\ c + d = 1, \end{cases}$$
that is, $c = -1$ and $d = 2$. So
$$t(0,1,0) = (-1,2)_F.$$
Finally, we find $e, f \in \mathbb{R}$ such that
$$t(0,0,1) = (0,0) = (e,f)_F.$$
Using the same method that we used to find c and d, we can take $e = f = 0$. So
$$t(0,0,1) = (0,0)_F.$$
The matrix of t with respect to the bases E and F is
$$\mathbf{A} = \begin{pmatrix} 1 & -1 & 0 \\ -1 & 2 & 0 \end{pmatrix}.$$

(c) We find the images of the vectors in the domain basis $E = \{(0,1,0),(1,1,1),(0,1,1)\}$:
$$t(0,1,0) = (0,1), \quad t(1,1,1) = (1,1),$$
$$t(0,1,1) = (0,1).$$
We find the F-coordinates of each of these image vectors, where $F = \{(1,3),(2,4)\}$.

We first find $a, b \in \mathbb{R}$ such that
$$t(0,1,0) = (0,1) = (a,b)_F.$$
Since
$$(a,b)_F = a(1,3) + b(2,4) = (a+2b, 3a+4b),$$
it follows that a and b are the solutions of
$$\begin{cases} a + 2b = 0, \\ 3a + 4b = 1, \end{cases}$$
that is, $a = 1$ and $b = -\frac{1}{2}$. So
$$t(0,1,0) = \left(1, -\tfrac{1}{2}\right)_F.$$
We next find $c, d \in \mathbb{R}$ such that
$$t(1,1,1) = (1,1) = (c,d)_F.$$
Since
$$(c,d)_F = c(1,3) + d(2,4) = (c+2d, 3c+4d),$$
it follows that c and d are the solutions of
$$\begin{cases} c + 2d = 1, \\ 3c + 4d = 1, \end{cases}$$
that is, $c = -1$ and $d = 1$. So
$$t(1,1,1) = (-1,1)_F.$$
Since $t(0,1,1) = (0,1) = t(0,1,0)$, we have
$$t(0,1,1) = \left(1, -\tfrac{1}{2}\right)_F.$$
The matrix of t with respect to the bases E and F is
$$\mathbf{A} = \begin{pmatrix} 1 & -1 & 1 \\ -\frac{1}{2} & 1 & -\frac{1}{2} \end{pmatrix}.$$

2.6 We follow Strategy 2.1.

(a) We find the images of the polynomials in the domain basis $E = \{1, x, x^2\}$:
$$t(1) = 0, \quad t(x) = 1, \quad t(x^2) = 2x.$$
We find the F-coordinates of each of these image vectors, where $F = \{2x, 1+x\}$.

We have
$$t(1) = 0 = (0,0)_F.$$
We next find $a, b \in \mathbb{R}$ such that
$$t(x) = 1 = (a,b)_F.$$
Since
$$(a,b)_F = a \times (2x) + b \times (1+x)$$
$$= b + (2a+b)x,$$
it follows that a and b are the solutions of
$$\begin{cases} b = 1, \\ 2a + b = 0, \end{cases}$$
that is, $a = -\frac{1}{2}$ and $b = 1$. So
$$t(x) = \left(-\tfrac{1}{2}, 1\right)_F.$$

Finally, using the same method that we used to find a and b, we have

$$t(x^2) = 2x = (1,0)_F.$$

The matrix of t with respect to the bases E and F is

$$\mathbf{A} = \begin{pmatrix} 0 & -\frac{1}{2} & 1 \\ 0 & 1 & 0 \end{pmatrix}.$$

The matrix representation of t with respect to the bases E and F is

$$\begin{pmatrix} a \\ b \\ c \end{pmatrix}_E \longmapsto \begin{pmatrix} 0 & -\frac{1}{2} & 1 \\ 0 & 1 & 0 \end{pmatrix} \begin{pmatrix} a \\ b \\ c \end{pmatrix}_E$$
$$= \begin{pmatrix} -\frac{1}{2}b + c \\ b \end{pmatrix}_F.$$

(b) We find the images of the polynomials in the domain basis $E = \{x, x^2, 1\}$:

$$t(x) = 1, \quad t(x^2) = 2x, \quad t(1) = 0.$$

We find the F-coordinates of each of these image vectors, where $F = \{2x, 1 + x\}$.

We know from part (a) that

$$t(x) = \left(-\tfrac{1}{2}, 1\right)_F, \quad t(x^2) = (1,0)_F,$$
$$t(1) = (0,0)_F.$$

The matrix of t with respect to the bases E and F is

$$\mathbf{A} = \begin{pmatrix} -\frac{1}{2} & 1 & 0 \\ 1 & 0 & 0 \end{pmatrix}.$$

The matrix representation of t with respect to the bases E and F is

$$\begin{pmatrix} a \\ b \\ c \end{pmatrix}_E \longmapsto \begin{pmatrix} -\frac{1}{2} & 1 & 0 \\ 1 & 0 & 0 \end{pmatrix} \begin{pmatrix} a \\ b \\ c \end{pmatrix}_E$$
$$= \begin{pmatrix} -\frac{1}{2}a + b \\ a \end{pmatrix}_F.$$

2.7 The functions in parts (a) and (d) are linear transformations, since they are of the form

$$t : \mathbb{R}^2 \longrightarrow \mathbb{R}^2$$
$$(x, y) \longmapsto (ax + by, cx + dy),$$

for some $a, b, c, d \in \mathbb{R}$.

The functions in parts (b) and (c) are not linear transformations, since they are not of this form.

2.8 We follow Strategy 2.1.

(a) First we find the images of the vectors in the domain basis $E = \{(1,0), (0,1)\}$:

$$t(1,0) = (3,2), \quad t(0,1) = (1,-1).$$

Next we find the F-coordinates of each of these image vectors, where $F = \{(1,0), (0,1)\}$:

$$t(1,0) = (3,2)_F, \quad t(0,1) = (1,-1)_F.$$

So the matrix of t with respect to the standard bases for the domain and codomain is

$$\mathbf{A} = \begin{pmatrix} 3 & 1 \\ 2 & -1 \end{pmatrix}.$$

(b) First we find the images of the vectors in the domain basis $E = \{(1,0,0), (0,1,0), (0,0,1)\}$:

$$t(1,0,0) = (1,1), \quad t(0,1,0) = (1,0),$$
$$t(0,0,1) = (0,-1).$$

Next we find the F-coordinates of each of these image vectors, where $F = \{(1,0), (0,1)\}$:

$$t(1,0,0) = (1,1)_F, \quad t(0,1,0) = (1,0)_F,$$
$$t(0,0,1) = (0,-1)_F.$$

So the matrix of t with respect to the standard bases for the domain and codomain is

$$\mathbf{A} = \begin{pmatrix} 1 & 1 & 0 \\ 1 & 0 & -1 \end{pmatrix}.$$

2.9 We follow Strategy 2.1.

(a) First we find the images of the vectors in the domain basis $E = \{(1,1), (0,1)\}$:

$$t(1,1) = (2,5), \quad t(0,1) = (1,3).$$

Next we find the F-coordinates of each of these image vectors, where $F = \{(1,0), (0,1)\}$:

$$t(1,1) = (2,5)_F, \quad t(0,1) = (1,3)_F.$$

So the matrix of t with respect to the bases E and F is

$$\mathbf{A} = \begin{pmatrix} 2 & 1 \\ 5 & 3 \end{pmatrix}.$$

Hence the matrix representation of t with respect to the bases E and F is

$$\begin{pmatrix} v_1 \\ v_2 \end{pmatrix}_E \longmapsto \begin{pmatrix} 2 & 1 \\ 5 & 3 \end{pmatrix} \begin{pmatrix} v_1 \\ v_2 \end{pmatrix}_E = \begin{pmatrix} 2v_1 + v_2 \\ 5v_1 + 3v_2 \end{pmatrix}_F.$$

(We have used v_1 and v_2 instead of x and y to emphasise that these are the coordinates with respect to a non-standard basis.)

(b) First we find the images of the vectors in the domain basis $E = \{(1,0), (0,1)\}$:

$$t(1,0) = (1,2), \quad t(0,1) = (1,3).$$

Next we find the F-coordinates of each of these image vectors, where $F = \{(2,1), (1,3)\}$.

We first find $a, b \in \mathbb{R}$ such that

$$t(1,0) = (1,2) = (a,b)_F.$$

Since

$$(a,b)_F = a(2,1) + b(1,3) = (2a + b, a + 3b),$$

it follows that a and b are the solutions of

$$\begin{cases} 2a + b = 1, \\ a + 3b = 2, \end{cases}$$

that is, $a = \frac{1}{5}$ and $b = \frac{3}{5}$. So

$$t(1,0) = \left(\tfrac{1}{5}, \tfrac{3}{5}\right)_F.$$

Similarly,

$$t(0,1) = (1,3) = (0,1)_F.$$

So the matrix of t with respect to the bases E and F is

$$\mathbf{A} = \begin{pmatrix} \frac{1}{5} & 0 \\ \frac{3}{5} & 1 \end{pmatrix}.$$

Hence the matrix representation of t with respect to the bases E and F is

$$\begin{pmatrix} x \\ y \end{pmatrix}_E \longmapsto \begin{pmatrix} \frac{1}{5} & 0 \\ \frac{3}{5} & 1 \end{pmatrix} \begin{pmatrix} x \\ y \end{pmatrix}_E = \begin{pmatrix} \frac{1}{5}x \\ \frac{3}{5}x + y \end{pmatrix}_F .$$

(c) First we find the images of the vectors in the domain basis $E = \{(1,2),(1,1)\}$:
$$t(1,2) = (3,8), \quad t(1,1) = (2,5).$$

Next we find the F-coordinates of each of these image vectors, where $F = \{(1,0),(3,1)\}$.

We first find $a,b \in \mathbb{R}$ such that
$$t(1,2) = (3,8) = (a,b)_F .$$

Since
$$(a,b)_F = a(1,0) + b(3,1) = (a+3b, b),$$
it follows that a and b are the solutions of
$$\begin{cases} a + 3b = 3, \\ \quad\quad b = 8, \end{cases}$$
that is, $a = -21$ and $b = 8$. So
$$t(1,2) = (-21,8)_F .$$

We next find $c,d \in \mathbb{R}$ such that
$$t(1,1) = (2,5) = (c,d)_F .$$

Since
$$(c,d)_F = c(1,0) + d(3,1) = (c+3d, d),$$
it follows that c and d are the solutions of
$$\begin{cases} c + 3d = 2, \\ \quad\quad d = 5, \end{cases}$$
that is, $c = -13$ and $d = 5$. So
$$t(1,1) = (-13,5)_F .$$

So the matrix of t with respect to the bases E and F is
$$\mathbf{A} = \begin{pmatrix} -21 & -13 \\ 8 & 5 \end{pmatrix} .$$

Hence the matrix representation of t with respect to the bases E and F is
$$\begin{pmatrix} v_1 \\ v_2 \end{pmatrix}_E \longmapsto \begin{pmatrix} -21 & -13 \\ 8 & 5 \end{pmatrix} \begin{pmatrix} v_1 \\ v_2 \end{pmatrix}_E$$
$$= \begin{pmatrix} -21v_1 - 13v_2 \\ 8v_1 + 5v_2 \end{pmatrix}_F .$$

2.10 We follow Strategy 2.1.

First we find the images of the vectors in the domain basis $E = \{e^x \cos x, e^x \sin x\}$:
$$t(e^x \cos x) = e^x \cos x - e^x \sin x,$$
$$t(e^x \sin x) = e^x \sin x + e^x \cos x.$$

Next we find the F-coordinates of each of these image vectors, where $F = \{e^x \cos x, e^x \sin x\}$:
$$t(e^x \cos x) = (1,-1)_F, \quad t(e^x \sin x) = (1,1)_F .$$

So the matrix of t with respect to the bases E and F is
$$\mathbf{A} = \begin{pmatrix} 1 & 1 \\ -1 & 1 \end{pmatrix} .$$

2.11 We follow Strategy 2.1.

(a) First we find the images of the vectors in the domain basis $E = \{1, x, x^2\}$:
$$t(1) = 1 + 0 = 1, \quad t(x) = x + 1,$$
$$t(x^2) = x^2 + 2x.$$

Next we find the F-coordinates of each of these image vectors, where $F = \{1, x, x^2\}$:
$$t(1) = (1,0,0)_F, \quad t(x) = (1,1,0)_F,$$
$$t(x^2) = (0,2,1)_F .$$

So the matrix of t with respect to the standard bases for the domain and codomain is
$$\mathbf{A} = \begin{pmatrix} 1 & 1 & 0 \\ 0 & 1 & 2 \\ 0 & 0 & 1 \end{pmatrix} .$$

Hence the matrix representation of t with respect to the standard bases for the domain and codomain is
$$\begin{pmatrix} a \\ b \\ c \end{pmatrix}_E \longmapsto \begin{pmatrix} 1 & 1 & 0 \\ 0 & 1 & 2 \\ 0 & 0 & 1 \end{pmatrix} \begin{pmatrix} a \\ b \\ c \end{pmatrix}_E = \begin{pmatrix} a+b \\ b+2c \\ c \end{pmatrix}_F .$$

(b) First we find the images of the vectors in the domain basis $E = \{1+x, x, x^2\}$:
$$t(1+x) = 1 + x + 1 = 2 + x,$$
$$t(x) = x + 1, \quad t(x^2) = x^2 + 2x.$$

Next we find the F-coordinates of each of these image vectors, where $F = \{1, x, x^2\}$:
$$t(1+x) = (2,1,0)_F, \quad t(x) = (1,1,0)_F,$$
$$t(x^2) = (0,2,1)_F .$$

So the matrix of t with respect to the bases E and F is
$$\mathbf{A} = \begin{pmatrix} 2 & 1 & 0 \\ 1 & 1 & 2 \\ 0 & 0 & 1 \end{pmatrix} .$$

Hence the matrix representation of t with respect to the bases E and F is
$$\begin{pmatrix} a \\ b \\ c \end{pmatrix}_E \longmapsto \begin{pmatrix} 2 & 1 & 0 \\ 1 & 1 & 2 \\ 0 & 0 & 1 \end{pmatrix} \begin{pmatrix} a \\ b \\ c \end{pmatrix}_E = \begin{pmatrix} 2a+b \\ a+b+2c \\ c \end{pmatrix}_F .$$

(c) First we find the images of the vectors in the domain basis $E = \{1, x, x^2\}$:
$$t(1) = 1 + 0 = 1, \quad t(x) = x + 1,$$
$$t(x^2) = x^2 + 2x.$$

Next we find the F-coordinates of each of these image vectors, where $F = \{1+x, x, x^2\}$.

We first find $a,b,c \in \mathbb{R}$ such that
$$t(1) = 1 = (a,b,c)_F .$$

Since
$$(a,b,c)_F = a(1+x) + bx + cx^2$$
$$= a + (a+b)x + cx^2,$$
it follows that a, b and c are the solutions of
$$\begin{cases} a \quad\quad\quad = 1, \\ a+b \quad\quad = 0, \\ \quad\quad\quad c = 0, \end{cases}$$

that is, $a = 1$, $b = -1$ and $c = 0$. So

$$t(1) = (1, -1, 0)_F.$$

Similarly,

$$t(x) = (1, 0, 0)_F, \quad t(x^2) = (0, 2, 1)_F.$$

So the matrix of t with respect to the bases E and F is

$$\mathbf{A} = \begin{pmatrix} 1 & 1 & 0 \\ -1 & 0 & 2 \\ 0 & 0 & 1 \end{pmatrix}.$$

Hence the matrix representation of t with respect to the bases E and F is

$$\begin{pmatrix} a \\ b \\ c \end{pmatrix}_E \longmapsto \begin{pmatrix} 1 & 1 & 0 \\ -1 & 0 & 2 \\ 0 & 0 & 1 \end{pmatrix} \begin{pmatrix} a \\ b \\ c \end{pmatrix}_E = \begin{pmatrix} a+b \\ -a+2c \\ c \end{pmatrix}_F.$$

3.1 (a) We have

$$\begin{aligned} s(t(x,y)) &= s(3x + y, -x) \\ &= (3x + y, (3x + y) - x) \\ &= (3x + y, 2x + y). \end{aligned}$$

Thus $s \circ t$ is given by

$$s \circ t : \mathbb{R}^2 \longrightarrow \mathbb{R}^2$$
$$(x, y) \longmapsto (3x + y, 2x + y).$$

(b) We have

$$\begin{aligned} t(s(x,y)) &= t(x, x + y) \\ &= (3x + (x + y), -x) \\ &= (4x + y, -x). \end{aligned}$$

Thus $t \circ s$ is given by

$$t \circ s : \mathbb{R}^2 \longrightarrow \mathbb{R}^2$$
$$(x, y) \longmapsto (4x + y, -x).$$

3.2 (a) It follows from the Composition Rule that the matrix of $s \circ t$ with respect to the standard bases for the domain and codomain is

$$\begin{pmatrix} 1 & 2 \\ 4 & 3 \end{pmatrix} \begin{pmatrix} 2 & 1 & 0 \\ 0 & 1 & 3 \end{pmatrix} = \begin{pmatrix} 2 & 3 & 6 \\ 8 & 7 & 9 \end{pmatrix}.$$

Thus the matrix representation of $s \circ t$ with respect to the standard bases for the domain and codomain is

$$s \circ t : \mathbb{R}^3 \longrightarrow \mathbb{R}^2$$
$$\begin{pmatrix} x \\ y \\ z \end{pmatrix} \longmapsto \begin{pmatrix} 2 & 3 & 6 \\ 8 & 7 & 9 \end{pmatrix} \begin{pmatrix} x \\ y \\ z \end{pmatrix}$$
$$= \begin{pmatrix} 2x + 3y + 6z \\ 8x + 7y + 9z \end{pmatrix}.$$

(b) It follows from the Composition Rule that the matrix of $s \circ t$ with respect to the standard bases for the domain and codomain is

$$\begin{pmatrix} 2 & 1 \\ 0 & 2 \\ 1 & 0 \end{pmatrix} \begin{pmatrix} 1 & 0 & 2 & 4 \\ 2 & 1 & 0 & 3 \end{pmatrix} = \begin{pmatrix} 4 & 1 & 4 & 11 \\ 4 & 2 & 0 & 6 \\ 1 & 0 & 2 & 4 \end{pmatrix}.$$

Thus the matrix representation of $s \circ t$ with respect to the standard bases for the domain and codomain is

$$s \circ t : \mathbb{R}^4 \longrightarrow \mathbb{R}^3$$
$$\begin{pmatrix} x \\ y \\ z \\ w \end{pmatrix} \longmapsto \begin{pmatrix} 4 & 1 & 4 & 11 \\ 4 & 2 & 0 & 6 \\ 1 & 0 & 2 & 4 \end{pmatrix} \begin{pmatrix} x \\ y \\ z \\ w \end{pmatrix}$$
$$= \begin{pmatrix} 4x + y + 4z + 11w \\ 4x + 2y + 6w \\ x + 2z + 4w \end{pmatrix}.$$

3.3 It follows from the Composition Rule that the matrix of $s \circ t$ with respect to the standard bases for the domain and codomain is

$$\begin{pmatrix} 0 & 1 & 0 \\ 0 & 0 & 2 \end{pmatrix} \begin{pmatrix} 2 & 2 & 4 \\ 0 & 1 & 0 \\ 0 & 0 & 1 \end{pmatrix} = \begin{pmatrix} 0 & 1 & 0 \\ 0 & 0 & 2 \end{pmatrix}.$$

Thus the matrix representation of $s \circ t$ with respect to the standard bases for the domain and codomain is

$$s \circ t : P_3 \longrightarrow P_3$$
$$\begin{pmatrix} a \\ b \\ c \end{pmatrix}_E \longmapsto \begin{pmatrix} 0 & 1 & 0 \\ 0 & 0 & 2 \end{pmatrix} \begin{pmatrix} a \\ b \\ c \end{pmatrix}_E = \begin{pmatrix} b \\ 2c \end{pmatrix}_F.$$

As expected, this is the same as the matrix representation for s.

3.4 Since

$$\begin{aligned} &s(t(x,y)) \\ &= s(4x - y, -3x + y) \\ &= ((4x - y) + (-3x + y), 3(4x - y) + 4(-3x + y)) \\ &= (x, y) \end{aligned}$$

and

$$\begin{aligned} &t(s(x,y)) \\ &= t(x + y, 3x + 4y) \\ &= (4(x + y) - (3x + 4y), -3(x + y) + (3x + 4y)) \\ &= (x, y), \end{aligned}$$

for each vector (x, y) in $\mathbb{R}^2$, s is the inverse function of t.

3.5 (a) Since t is a linear transformation between two vector spaces of the same dimension, we use Strategy 3.1.

First we find a matrix representation of t. Since

$$t(1, 0) = (2, 4), \quad t(0, 1) = (1, 2),$$

the matrix representation of t with respect to the standard bases for the domain and codomain is

$$\begin{pmatrix} x \\ y \end{pmatrix} \longmapsto \begin{pmatrix} 2 & 1 \\ 4 & 2 \end{pmatrix} \begin{pmatrix} x \\ y \end{pmatrix} = \begin{pmatrix} 2x + y \\ 4x + 2y \end{pmatrix}.$$

Next we evaluate the determinant of the matrix
$$\mathbf{A} = \begin{pmatrix} 2 & 1 \\ 4 & 2 \end{pmatrix}.$$
We have
$$\det \mathbf{A} = \begin{vmatrix} 2 & 1 \\ 4 & 2 \end{vmatrix} = 4 - 4 = 0.$$
Since $\det \mathbf{A} = 0$, t is not invertible.

(b) Since t is a linear transformation between two vector spaces of the same dimension, we use Strategy 3.1.

First we find a matrix representation of t. Since
$$t(1,0) = (1,3), \quad t(0,1) = (-1,1),$$
the matrix representation of t with respect to the standard bases for the domain and codomain is
$$\begin{pmatrix} x \\ y \end{pmatrix} \longmapsto \begin{pmatrix} 1 & -1 \\ 3 & 1 \end{pmatrix} \begin{pmatrix} x \\ y \end{pmatrix} = \begin{pmatrix} x - y \\ 3x + y \end{pmatrix}.$$
Next we evaluate the determinant of the matrix
$$\mathbf{A} = \begin{pmatrix} 1 & -1 \\ 3 & 1 \end{pmatrix}.$$
We have
$$\det \mathbf{A} = \begin{vmatrix} 1 & -1 \\ 3 & 1 \end{vmatrix} = 1 - (-3) = 4.$$
Since $\det \mathbf{A} \ne 0$, t is invertible.

We now find the inverse function of t, $t^{-1} : \mathbb{R}^2 \longrightarrow \mathbb{R}^2$. According to Strategy 3.1, t^{-1} has the matrix representation
$$\mathbf{v} \longmapsto \mathbf{A}^{-1} \mathbf{v}$$
with respect to the standard bases for the domain and codomain. Since
$$\mathbf{A}^{-1} = \tfrac{1}{4} \begin{pmatrix} 1 & 1 \\ -3 & 1 \end{pmatrix} = \begin{pmatrix} \frac{1}{4} & \frac{1}{4} \\ -\frac{3}{4} & \frac{1}{4} \end{pmatrix},$$
it follows that t^{-1} has the matrix representation
$$\begin{pmatrix} x \\ y \end{pmatrix} \longmapsto \begin{pmatrix} \frac{1}{4} & \frac{1}{4} \\ -\frac{3}{4} & \frac{1}{4} \end{pmatrix} \begin{pmatrix} x \\ y \end{pmatrix} = \begin{pmatrix} \frac{1}{4}x + \frac{1}{4}y \\ -\frac{3}{4}x + \frac{1}{4}y \end{pmatrix}.$$
So t^{-1} is the linear transformation
$$t^{-1} : \mathbb{R}^2 \longrightarrow \mathbb{R}^2$$
$$(x, y) \longmapsto \left(\tfrac{1}{4}x + \tfrac{1}{4}y, -\tfrac{3}{4}x + \tfrac{1}{4}y \right).$$

(c) Since t is a linear transformation between two vector spaces of the same dimension, we use Strategy 3.1.

First we find a matrix representation of t. Since
$$t(1,0,0) = (2,-1,0), \quad t(0,1,0) = (0,3,0)$$
$$t(0,0,1) = (0,0,1),$$
the matrix representation of t with respect to the standard bases for the domain and codomain is
$$\begin{pmatrix} x \\ y \\ z \end{pmatrix} \longmapsto \begin{pmatrix} 2 & 0 & 0 \\ -1 & 3 & 0 \\ 0 & 0 & 1 \end{pmatrix} \begin{pmatrix} x \\ y \\ z \end{pmatrix} = \begin{pmatrix} 2x \\ 3y - x \\ z \end{pmatrix}.$$

Next we evaluate the determinant of the matrix
$$\mathbf{A} = \begin{pmatrix} 2 & 0 & 0 \\ -1 & 3 & 0 \\ 0 & 0 & 1 \end{pmatrix}.$$
We have
$$\det \mathbf{A} = \begin{vmatrix} 2 & 0 & 0 \\ -1 & 3 & 0 \\ 0 & 0 & 1 \end{vmatrix} = 2 \begin{vmatrix} 3 & 0 \\ 0 & 1 \end{vmatrix} - 0 + 0$$
$$= 2 \times 3 = 6.$$
Since $\det \mathbf{A} \ne 0$, t is invertible.

We now find the inverse function of t, $t^{-1} : \mathbb{R}^3 \longrightarrow \mathbb{R}^3$. According to Strategy 3.1, t^{-1} has the matrix representation
$$\mathbf{v} \longmapsto \mathbf{A}^{-1} \mathbf{v}$$
with respect to the standard bases for the domain and codomain.

Using row-reduction, we find
$$\mathbf{A}^{-1} = \begin{pmatrix} \frac{1}{2} & 0 & 0 \\ \frac{1}{6} & \frac{1}{3} & 0 \\ 0 & 0 & 1 \end{pmatrix},$$
so t^{-1} has the matrix representation
$$\begin{pmatrix} x \\ y \\ z \end{pmatrix} \longmapsto \begin{pmatrix} \frac{1}{2} & 0 & 0 \\ \frac{1}{6} & \frac{1}{3} & 0 \\ 0 & 0 & 1 \end{pmatrix} \begin{pmatrix} x \\ y \\ z \end{pmatrix} = \begin{pmatrix} \frac{1}{2}x \\ \frac{1}{6}x + \frac{1}{3}y \\ z \end{pmatrix}.$$
So t^{-1} is the linear transformation
$$t^{-1} : \mathbb{R}^3 \longrightarrow \mathbb{R}^3$$
$$(x, y, z) \longmapsto \left(\tfrac{1}{2}x, \tfrac{1}{6}x + \tfrac{1}{3}y, z \right).$$

(d) Since t is a linear transformation between two vector spaces of different dimensions, it follows from the corollary to Theorem 3.2 that t is not invertible.

3.6 The linear transformation
$$t : P_3 \longrightarrow \mathbb{R}^3$$
$$a + bx + cx^2 \longmapsto (a, b, c)$$
is one-one and onto and hence invertible. It is therefore an isomorphism.

3.7 The vector spaces $\mathbb{R}^2$, $\mathbb{C}$ and P_2 are isomorphic, since they are all two-dimensional.

The vector spaces $\mathbb{R}^3$ and P_3 are isomorphic, since they are both three-dimensional.

3.8 (a) We have
$$s(t(x,y)) = s(x+y, 2y)$$
$$= (3(x+y), 2(2y))$$
$$= (3x + 3y, 4y).$$
Thus
$$s \circ t : \mathbb{R}^2 \longrightarrow \mathbb{R}^2$$
$$(x, y) \longmapsto (3x + 3y, 4y).$$

(b) We have
$$t(s(x, y)) = t(3x, 2y)$$
$$= (3x + 2y, 2(2y))$$
$$= (3x + 2y, 4y).$$
Thus
$$t \circ s : \mathbb{R}^2 \longrightarrow \mathbb{R}^2$$
$$(x, y) \longmapsto (3x + 2y, 4y).$$

(c) We have
$$t(t(x, y)) = t(x + y, 2y)$$
$$= ((x + y) + 2y, 2(2y))$$
$$= (x + 3y, 4y).$$
Thus
$$t \circ t : \mathbb{R}^2 \longrightarrow \mathbb{R}^2$$
$$(x, y) \longmapsto (x + 3y, 4y).$$

3.9 It follows from the Composition Rule that the matrix of $s \circ t$ with respect to the standard bases for the domain and codomain is
$$\begin{pmatrix} 3 & 1 \\ 2 & -1 \end{pmatrix} \begin{pmatrix} 1 & 1 & 0 \\ 1 & 0 & -1 \end{pmatrix} = \begin{pmatrix} 4 & 3 & -1 \\ 1 & 2 & 1 \end{pmatrix}.$$
Thus the matrix representation of $s \circ t$ with respect to the standard bases for the domain and codomain is
$$s \circ t : \mathbb{R}^3 \longrightarrow \mathbb{R}^2$$
$$\begin{pmatrix} x \\ y \\ z \end{pmatrix} \longmapsto \begin{pmatrix} 4 & 3 & -1 \\ 1 & 2 & 1 \end{pmatrix} \begin{pmatrix} x \\ y \\ z \end{pmatrix} = \begin{pmatrix} 4x + 3y - z \\ x + 2y + z \end{pmatrix}.$$

3.10 It follows from the Composition Rule that the matrix of $t \circ t$ with respect to the basis $E = \{e^x \cos x, e^x \sin x\}$ for both the domain and codomain is
$$\begin{pmatrix} 1 & 1 \\ -1 & 1 \end{pmatrix} \begin{pmatrix} 1 & 1 \\ -1 & 1 \end{pmatrix} = \begin{pmatrix} 0 & 2 \\ -2 & 0 \end{pmatrix}.$$
Thus the matrix representation of $t \circ t$ with respect to the basis $E = \{e^x \cos x, e^x \sin x\}$ for both the domain and codomain is
$$t \circ t : V \longrightarrow V$$
$$\begin{pmatrix} a \\ b \end{pmatrix}_E \longmapsto \begin{pmatrix} 0 & 2 \\ -2 & 0 \end{pmatrix} \begin{pmatrix} a \\ b \end{pmatrix}_E = \begin{pmatrix} 2b \\ -2a \end{pmatrix}_E.$$
So the function $f(x) = ae^x \cos x + be^x \sin x$ in V maps to the function $g(x) = 2be^x \cos x - 2ae^x \sin x$ in V under the linear transformation $t \circ t$.

Remark We could find this directly by differentiating $ae^x \cos x + be^x \sin x$ twice. The matrix multiplication is, however, very simple and is possibly easier than differentiating. To obtain higher derivatives, we could repeatedly apply the Composition Rule.

3.11 (a) Since t is a linear transformation between two vector spaces of the same dimension, we use Strategy 3.1.

First we find a matrix representation of t. Since
$$t(1, 0) = (3, 0), \quad t(0, 1) = (1, 1),$$
the matrix representation of t with respect to the standard bases for the domain and codomain is
$$\begin{pmatrix} x \\ y \end{pmatrix} \longmapsto \begin{pmatrix} 3 & 1 \\ 0 & 1 \end{pmatrix} \begin{pmatrix} x \\ y \end{pmatrix} = \begin{pmatrix} 3x + y \\ y \end{pmatrix}.$$
Next we evaluate the determinant of the matrix
$$\mathbf{A} = \begin{pmatrix} 3 & 1 \\ 0 & 1 \end{pmatrix}.$$
We have
$$\det \mathbf{A} = \begin{vmatrix} 3 & 1 \\ 0 & 1 \end{vmatrix} = 3 - 0 = 3.$$
Since $\det \mathbf{A} \neq 0$, t is invertible.

We now find the inverse function of t, $t^{-1} : \mathbb{R}^2 \longrightarrow \mathbb{R}^2$. According to Strategy 3.1, t^{-1} has the matrix representation
$$\mathbf{v} \longmapsto \mathbf{A}^{-1} \mathbf{v}$$
with respect to the standard bases for the domain and codomain. Since
$$\mathbf{A}^{-1} = \tfrac{1}{3} \begin{pmatrix} 1 & -1 \\ 0 & 3 \end{pmatrix} = \begin{pmatrix} \frac{1}{3} & -\frac{1}{3} \\ 0 & 1 \end{pmatrix},$$
it follows that t^{-1} has the matrix representation
$$\begin{pmatrix} x \\ y \end{pmatrix} \longmapsto \begin{pmatrix} \frac{1}{3} & -\frac{1}{3} \\ 0 & 1 \end{pmatrix} \begin{pmatrix} x \\ y \end{pmatrix} = \begin{pmatrix} \frac{1}{3}x - \frac{1}{3}y \\ y \end{pmatrix}.$$
So t^{-1} is the linear transformation
$$t^{-1} : \mathbb{R}^2 \longrightarrow \mathbb{R}^2$$
$$(x, y) \longmapsto (\tfrac{1}{3}x - \tfrac{1}{3}y, y).$$

(b) Since t is a linear transformation between two vector spaces of the same dimension, we use Strategy 3.1.

First we find a matrix representation of t. Since
$$t(1, 0) = (2, 1), \quad t(0, 1) = (6, 3),$$
the matrix representation of t with respect to the standard bases for the domain and codomain is
$$\begin{pmatrix} x \\ y \end{pmatrix} \longmapsto \begin{pmatrix} 2 & 6 \\ 1 & 3 \end{pmatrix} \begin{pmatrix} x \\ y \end{pmatrix} = \begin{pmatrix} 2x + 6y \\ x + 3y \end{pmatrix}.$$
Next we evaluate the determinant of the matrix
$$\mathbf{A} = \begin{pmatrix} 2 & 6 \\ 1 & 3 \end{pmatrix}.$$
We have
$$\det \mathbf{A} = \begin{vmatrix} 2 & 6 \\ 1 & 3 \end{vmatrix} = 6 - 6 = 0.$$
Since $\det \mathbf{A} = 0$, t is not invertible.

(c) Since t is a linear transformation between two vector spaces of different dimensions, it follows from the corollary to Theorem 3.2 that t is not invertible.

(d) Since t is a linear transformation between two vector spaces of the same dimension, we use Strategy 3.1.

First we find a matrix representation of t. Since

$$t(1,0,0) = (1,2,0), \quad t(0,1,0) = (0,1,2),$$
$$t(0,0,1) = (1,3,1),$$

the matrix representation of t with respect to the standard bases for the domain and codomain is

$$\begin{pmatrix} x \\ y \\ z \end{pmatrix} \longmapsto \begin{pmatrix} 1 & 0 & 1 \\ 2 & 1 & 3 \\ 0 & 2 & 1 \end{pmatrix} \begin{pmatrix} x \\ y \\ z \end{pmatrix} = \begin{pmatrix} x + z \\ 2x + y + 3z \\ 2y + z \end{pmatrix}.$$

Next we evaluate the determinant of the matrix

$$\mathbf{A} = \begin{pmatrix} 1 & 0 & 1 \\ 2 & 1 & 3 \\ 0 & 2 & 1 \end{pmatrix}.$$

We have

$$\det \mathbf{A} = \begin{vmatrix} 1 & 0 & 1 \\ 2 & 1 & 3 \\ 0 & 2 & 1 \end{vmatrix} = 1 \begin{vmatrix} 1 & 3 \\ 2 & 1 \end{vmatrix} - 0 + 1 \begin{vmatrix} 2 & 1 \\ 0 & 2 \end{vmatrix}$$
$$= 1(1 - 6) + 1(4 - 0)$$
$$= -5 + 4 = -1.$$

Since $\det \mathbf{A} \neq 0$, t is invertible.

We now find the inverse function of t.

According to Strategy 3.1, t^{-1} has the matrix representation

$$\mathbf{v} \longmapsto \mathbf{A}^{-1}\mathbf{v}$$

with respect to the standard bases for the domain and codomain.

Using row-reduction, we find that

$$\mathbf{A}^{-1} = \begin{pmatrix} 5 & -2 & 1 \\ 2 & -1 & 1 \\ -4 & 2 & -1 \end{pmatrix},$$

so t^{-1} has the matrix representation

$$\begin{pmatrix} x \\ y \\ z \end{pmatrix} \longmapsto \begin{pmatrix} 5 & -2 & 1 \\ 2 & -1 & 1 \\ -4 & 2 & -1 \end{pmatrix} \begin{pmatrix} x \\ y \\ z \end{pmatrix}$$
$$= \begin{pmatrix} 5x - 2y + z \\ 2x - y + z \\ -4x + 2y - z \end{pmatrix}.$$

So t^{-1} is the linear transformation

$$t^{-1} : \mathbb{R}^3 \longrightarrow \mathbb{R}^3$$
$$(x, y, z) \longmapsto (5x - 2y + z, 2x - y + z,$$
$$-4x + 2y - z).$$

3.12 Since t is a linear transformation between two vector spaces of the same dimension, we use Strategy 3.1.

First we find a matrix representation of t. We know from the solution to Exercise 2.11(a) that

$$\begin{pmatrix} a \\ b \\ c \end{pmatrix}_E \longmapsto \begin{pmatrix} 1 & 1 & 0 \\ 0 & 1 & 2 \\ 0 & 0 & 1 \end{pmatrix} \begin{pmatrix} a \\ b \\ c \end{pmatrix}_E = \begin{pmatrix} a + b \\ b + 2c \\ c \end{pmatrix}_F$$

is the matrix representation of t with respect to the standard basis $\{1, x, x^2\}$ for both the domain and codomain.

Next we evaluate the determinant of the matrix

$$\mathbf{A} = \begin{pmatrix} 1 & 1 & 0 \\ 0 & 1 & 2 \\ 0 & 0 & 1 \end{pmatrix}.$$

We have

$$\det \mathbf{A} = \begin{vmatrix} 1 & 1 & 0 \\ 0 & 1 & 2 \\ 0 & 0 & 1 \end{vmatrix} = 1 \begin{vmatrix} 1 & 2 \\ 0 & 1 \end{vmatrix} - 1 \begin{vmatrix} 0 & 2 \\ 0 & 1 \end{vmatrix} + 0$$
$$= 1(1 - 0) - 1(0 - 0) + 0$$
$$= 1.$$

Since $\det \mathbf{A} \neq 0$, t is invertible.

3.13 (a) An isomorphism is

$$t : V \longrightarrow \mathbb{R}^2$$
$$ae^x \cos x + be^x \sin x \longmapsto (a, b).$$

(The matrix of t with respect to the basis $E = \{e^x \cos x, e^x \sin x\}$ for V and the standard basis for $\mathbb{R}^2$ is $\mathbf{I}_2$. Since t has a matrix representation, it is a linear transformation, by Theorem 2.2. Since $\mathbf{I}_2$ is invertible, it follows from the Inverse Rule that t is an invertible linear transformation; that is, t is an isomorphism.)

(b) An isomorphism is

$$t : V \longrightarrow P_2$$
$$ae^x \cos x + be^x \sin x \longmapsto a + bx.$$

(The matrix of t with respect to the basis $E = \{e^x \cos x, e^x \sin x\}$ for V and the standard basis for P_2 is $\mathbf{I}_2$.)

3.14 We show that, under the operation of composition, the set G of invertible linear transformations from $\mathbb{R}^n$ to itself satisfies the four group axioms.

G1 CLOSURE

Let $t : \mathbb{R}^n \longrightarrow \mathbb{R}^n$ and $s : \mathbb{R}^n \longrightarrow \mathbb{R}^n$ be two invertible linear transformations in G. Let $\mathbf{A}$ be the matrix of t and let $\mathbf{B}$ be the matrix of s, with respect to the standard bases for the domain and codomain. Then it follows from the Composition Rule that $\mathbf{BA}$ is the matrix of the linear transformation $s \circ t : \mathbb{R}^n \longrightarrow \mathbb{R}^n$, with respect to the standard bases for the domain and codomain.

Since t and s are invertible, it follows from the Inverse Rule that $\mathbf{A}$ and $\mathbf{B}$ are invertible. Thus $\mathbf{BA}$ is invertible, since $(\mathbf{BA})^{-1} = \mathbf{A}^{-1}\mathbf{B}^{-1}$. It follows from the Inverse Rule that $s \circ t$ is invertible. Thus $s \circ t$ belongs to G; that is, G is closed under composition.

G2 IDENTITY

We claim that the identity transformation

$$i : \mathbb{R}^n \longrightarrow \mathbb{R}^n$$
$$\mathbf{v} \longmapsto \mathbf{v}$$

is the identity element of G. The matrix of i is $\mathbf{I}_n$, with respect to the standard bases for the domain and codomain.

Since $\mathbf{I}_n$ is invertible, it follows from the Inverse Rule that i is invertible; that is, i belongs to G.

Suppose that $t : \mathbb{R}^n \longrightarrow \mathbb{R}^n$ belongs to G and $\mathbf{A}$ is the matrix of t with respect to the standard bases for the domain and codomain. Since $\mathbf{A}\mathbf{I}_n = \mathbf{I}_n\mathbf{A} = \mathbf{A}$, it follows from the Composition Rule that $i \circ t = t \circ i = t$.

So the identity transformation is the identity element of G.

G3 INVERSES

Suppose that $t : \mathbb{R}^n \longrightarrow \mathbb{R}^n$ belongs to G. It follows from the Inverse Rule that the inverse function of t, $t^{-1} : \mathbb{R}^n \longrightarrow \mathbb{R}^n$, is also a linear transformation. It is invertible, since the inverse function of t^{-1} is just t. So the inverse of t belongs to G.

G4 ASSOCIATIVITY

We know that composition of functions is associative.

We have shown that, under the operation of composition, G satisfies the four group axioms and is therefore a group.

4.1 (a) The image of this linear transformation is the x-axis. This is a subspace of the codomain.

(b) The image of this linear transformation is the line $y = x$. This is a subspace of the codomain.

4.2 We have

$$t(1,0) = (1,1), \quad t(0,1) = (0,0).$$

We found in Exercise 4.1(b) that the image of t is the line $y = x$; that is,

$$\mathrm{Im}(t) = \{(k,k) : k \in \mathbb{R}\}.$$

Thus $\mathrm{Im}(t)$ is spanned by $(1,1) = t(1,0)$.

4.3 We follow the steps of Strategy 4.1.

(a) We take the standard basis $\{(1,0),(0,1)\}$ for the domain $\mathbb{R}^2$.

We determine the images of these basis vectors:

$$t(1,0) = (1,2), \quad t(0,1) = (0,1).$$

The set $\{(1,2),(0,1)\}$ is linearly independent, so it is a basis for $\mathrm{Im}(t)$.

Since the basis has two elements,

$$\dim \mathrm{Im}(t) = 2.$$

(b) We take the standard basis $\{(1,0,0),(0,1,0),(0,0,1)\}$ for the domain $\mathbb{R}^3$.

We determine the images of these basis vectors:

$$t(1,0,0) = (1,1,1), \quad t(0,1,0) = (2,0,1),$$
$$t(0,0,1) = (3,1,2).$$

The set $\{(1,1,1),(2,0,1),(3,1,2)\}$ is not linearly independent. In fact,

$$(3,1,2) = (1,1,1) + (2,0,1),$$

so we discard $(3,1,2)$ to give the set $\{(1,1,1),(2,0,1)\}$.

The vectors $(1,1,1)$ and $(2,0,1)$ are linearly independent, so $\{(1,1,1),(2,0,1)\}$ is a basis for $\mathrm{Im}(t)$.

Since the basis has two elements,

$$\dim \mathrm{Im}(t) = 2.$$

Remark You may have chosen to discard $(1,1,1)$ or $(2,0,1)$ instead. This would still give a correct answer.

(c) We take the standard basis $\{1, x, x^2\}$ for the domain P_3.

We determine the images of these basis vectors:

$$t(1) = 0, \quad t(x) = 1, \quad t(x^2) = 2x.$$

The set $\{0, 1, 2x\}$ is not linearly independent, since it contains the zero vector. We discard 0 to give the set $\{1, 2x\}$.

The vectors 1 and $2x$ are linearly independent, so $\{1, 2x\}$ is a basis for $\mathrm{Im}(t)$.

Since the basis has two elements,

$$\dim \mathrm{Im}(t) = 2.$$

4.4 (a) We know from Exercise 4.3(a) that $\dim \mathrm{Im}(t) = 2$. Thus $\mathrm{Im}(t)$ is the whole of the two-dimensional codomain $\mathbb{R}^2$, so t is onto.

(b) We know from Exercise 4.3(b) that $\dim \mathrm{Im}(t) = 2$. Thus $\mathrm{Im}(t)$ is not the whole of the three-dimensional codomain $\mathbb{R}^3$, so t is not onto.

(c) We know from Exercise 4.3(c) that $\dim \mathrm{Im}(t) = 2$. Thus $\mathrm{Im}(t)$ is the whole of the two-dimensional codomain P_2, so t is onto.

4.5 (a) For this linear transformation,

$$t(x,y,z) = \mathbf{0}$$

if and only if

$$(x,0) = (0,0),$$

that is, if and only if x, y and z satisfy the equation

$$x = 0.$$

Thus the kernel of t is the (y,z)-plane. This is a subspace of the domain $\mathbb{R}^3$.

(b) For this linear transformation,
$$t(x, y) = \mathbf{0}$$
if and only if
$$(x, x) = (0, 0),$$
that is, if and only if x and y satisfy the equation
$$x = 0.$$
Thus the kernel of t is the y-axis. This is a subspace of the domain $\mathbb{R}^2$.

4.6 (a) The kernel of t is the set of vectors (x, y) in $\mathbb{R}^2$ that satisfy
$$t(x, y) = \mathbf{0},$$
that is,
$$(x, 2x + y) = (0, 0).$$
We look for the values of x and y that satisfy the simultaneous equations
$$\begin{cases} x & = 0, \\ 2x + y = 0. \end{cases}$$
Putting $x = 0$ from the first equation into the second equation, we obtain $y = 0$.

So the kernel of t is
$$\mathrm{Ker}(t) = \{(0, 0)\}.$$
Since this contains only the zero vector,
$$\dim \mathrm{Ker}(t) = 0.$$

(b) The kernel of t is the set of vectors (x, y, z) in $\mathbb{R}^3$ that satisfy
$$t(x, y, z) = \mathbf{0},$$
that is,
$$(x + 2y + 3z, x + z, x + y + 2z) = (0, 0, 0).$$
We look for the values of x, y and z that satisfy the simultaneous equations
$$\begin{cases} x + 2y + 3z = 0, \\ x \quad\quad + \ z = 0, \\ x + \ y + 2z = 0. \end{cases}$$
To solve these equations using row-reduction, we use the augmented matrix.

$$\begin{array}{c} \mathbf{r}_1 \\ \mathbf{r}_2 \\ \mathbf{r}_3 \end{array} \quad \left(\begin{array}{ccc|c} 1 & 2 & 3 & 0 \\ 1 & 0 & 1 & 0 \\ 1 & 1 & 2 & 0 \end{array} \right)$$

$$\begin{array}{c} \mathbf{r}_2 \to \mathbf{r}_2 - \mathbf{r}_1 \\ \mathbf{r}_3 \to \mathbf{r}_3 - \mathbf{r}_1 \end{array} \quad \left(\begin{array}{ccc|c} 1 & 2 & 3 & 0 \\ 0 & -2 & -2 & 0 \\ 0 & -1 & -1 & 0 \end{array} \right)$$

$$\begin{array}{c} \mathbf{r}_2 \to -\frac{1}{2}\mathbf{r}_2 \end{array} \quad \left(\begin{array}{ccc|c} 1 & 2 & 3 & 0 \\ 0 & 1 & 1 & 0 \\ 0 & -1 & -1 & 0 \end{array} \right)$$

$$\begin{array}{c} \mathbf{r}_1 \to \mathbf{r}_1 - 2\mathbf{r}_2 \\ \\ \mathbf{r}_3 \to \mathbf{r}_3 + \mathbf{r}_2 \end{array} \quad \left(\begin{array}{ccc|c} 1 & 0 & 1 & 0 \\ 0 & 1 & 1 & 0 \\ 0 & 0 & 0 & 0 \end{array} \right)$$

The augmented matrix is now in row-reduced form. It gives
$$\begin{cases} x \quad\quad + z = 0, \\ \quad y + z = 0. \end{cases}$$
Assigning the parameter k to the unknown z, we obtain
$$x = -k, \quad y = -k, \quad z = k.$$
So the kernel of t is
$$\mathrm{Ker}(t) = \{(-k, -k, k) : k \in \mathbb{R}\},$$
that is, $\mathrm{Ker}(t)$ is the line through $(0, 0, 0)$ and $(-1, -1, 1)$.

Thus
$$\dim \mathrm{Ker}(t) = 1.$$

4.7 Let $p(x) = a + bx + cx^2$ be a polynomial in P_3. Then
$$t(p(x)) = b + 2cx.$$
The kernel of t is the set of polynomials $p(x) = a + bx + cx^2$ in P_3 that satisfy
$$t(p(x)) = \mathbf{0},$$
that is,
$$b + 2cx = 0.$$
We look for the values of a, b and c that satisfy the simultaneous equations
$$\begin{cases} b \quad\quad = 0, \\ \quad 2c = 0. \end{cases}$$
So a can take any real value, $b = 0$ and $c = 0$.

Thus the kernel of t is
$$\mathrm{Ker}(t) = \{p(x) : p(x) = a, \ a \in \mathbb{R}\},$$
that is, the set of constant polynomials.

It follows that
$$\dim \mathrm{Ker}(t) = 1.$$

4.8 For the linear transformation t in Exercise 4.6(a), $\mathrm{Ker}(t) = \{\mathbf{0}\}$. Thus t is one-one.

For the linear transformation t in Exercise 4.6(b), $\mathrm{Ker}(t) \neq \{\mathbf{0}\}$. Thus t is not one-one.

For the linear transformation t in Exercise 4.7, $\mathrm{Ker}(t) \neq \{\mathbf{0}\}$. Thus t is not one-one.

4.9 (a) For the linear transformation
$$t : \mathbb{R}^2 \longrightarrow \mathbb{R}^2$$
$$(x, y) \longmapsto (x, 2x + y),$$
we found in Exercise 4.3(a) that $\dim \mathrm{Im}(t) = 2$, and in Exercise 4.6(a) that $\dim \mathrm{Ker}(t) = 0$. Thus
$$\dim \mathrm{Im}(t) + \dim \mathrm{Ker}(t) = 2 + 0 = 2,$$
which is the dimension of the domain $\mathbb{R}^2$.

(b) For the linear transformation

$$t : \mathbb{R}^3 \longrightarrow \mathbb{R}^3$$
$$(x, y, z) \longmapsto (x + 2y + 3z, x + z, x + y + 2z),$$

we found in Exercise 4.3(b) that $\dim \mathrm{Im}(t) = 2$, and in Exercise 4.6(b) that $\dim \mathrm{Ker}(t) = 1$. Thus

$$\dim \mathrm{Im}(t) + \dim \mathrm{Ker}(t) = 2 + 1 = 3,$$

which is the dimension of the domain $\mathbb{R}^3$.

(c) For the linear transformation

$$t : P_3 \longrightarrow P_2$$
$$p(x) \longmapsto p'(x),$$

we found in Exercise 4.3(c) that $\dim \mathrm{Im}(t) = 2$, and in Exercise 4.7 that $\dim \mathrm{Ker}(t) = 1$. Thus

$$\dim \mathrm{Im}(t) + \dim \mathrm{Ker}(t) = 2 + 1 = 3,$$

which is the dimension of the domain P_3.

4.10 (a) In this case, the dimension of the codomain (which is 3) is greater than the dimension of the domain (which is 2), so t is not onto.

(b) In this case, the codomain and the domain both have dimension 2. There are two possibilities:

EITHER t is both one-one and onto;

OR t is neither one-one nor onto.

(c) In this case, the dimension of the codomain (which is 2) is less than the dimension of the domain (which is 3), so t is not one-one.

4.11 The number of solutions to this system of equations is the same as the number of vectors that map to $(1, 1, 1)$ under the linear transformation

$$t : \mathbb{R}^3 \longrightarrow \mathbb{R}^3$$
$$(x, y, z) \longmapsto (x + 2y + 3z, x + z, x + y + 2z).$$

We know from the solution to Exercise 4.3(b) that $(1, 1, 1)$ is in the image of t, and from Exercise 4.6(b) that $\mathrm{Ker}(t) \neq \{\mathbf{0}\}$. Thus the system of equations has infinitely many solutions.

4.12 (a) This is a system of two simultaneous linear equations in three unknowns. Since $3 > 2$, the system has either no solutions or infinitely many solutions.

(b) This is a system of three simultaneous linear equations in three unknowns. There are two possibilities:

EITHER the system has exactly one solution for each set of values of a, b and c;

OR there are some values of a, b and c for which the system has no solutions, and for all other values of a, b and c, the system has infinitely many solutions.

4.13 (a) Since the dimension of the codomain is less than the dimension of the domain, it follows from Theorem 4.5 that t is not one-one.

(b) The kernel of t is the set of vectors (x, y, z) in $\mathbb{R}^3$ that satisfy

$$t(x, y, z) = \mathbf{0},$$

that is,

$$(x + y, x - z) = (0, 0).$$

Thus we are looking for the values of x, y and z that satisfy the simultaneous equations

$$\begin{cases} x + y & = 0, \\ x & - z = 0. \end{cases}$$

Assigning the parameter k to the unknown z, we obtain

$$x = k, \quad y = -k, \quad z = k.$$

So the kernel of t is

$$\mathrm{Ker}(t) = \{(k, -k, k) : k \in \mathbb{R}\},$$

that is, $\mathrm{Ker}(t)$ is the line through $(0, 0, 0)$ and $(1, -1, 1)$. Thus

$$\dim \mathrm{Ker}(t) = 1.$$

(c) Since the dimension of the domain of t is 3, it follows from the Dimension Theorem that

$$\dim \mathrm{Im}(t) + \dim \mathrm{Ker}(t) = 3.$$

Since $\dim \mathrm{Ker}(t) = 1$, it follows that

$$\dim \mathrm{Im}(t) = 2.$$

Thus $\mathrm{Im}(t)$ is a two-dimensional subspace of the codomain $\mathbb{R}^2$ and is hence equal to $\mathbb{R}^2$.

(d) We saw in part (a) that t is not one-one; this was confirmed in part (b), where we saw that $\mathrm{Ker}(t) \neq \{\mathbf{0}\}$.

We saw in part (c) that $\mathrm{Im}(t)$ is the whole of the codomain $\mathbb{R}^2$; that is, t is onto.

4.14 (a) We follow Strategy 4.1.

We take the standard basis $\{(1, 0, 0), (0, 1, 0), (0, 0, 1)\}$ for the domain $\mathbb{R}^3$.

We determine the images of these basis vectors:

$$t(1, 0, 0) = (1, 0, 1), \quad t(0, 1, 0) = (2, 1, 0),$$
$$t(0, 0, 1) = (3, 1, 1).$$

The set $\{(1, 0, 1), (2, 1, 0), (3, 1, 1)\}$ is not linearly independent. In fact,

$$(3, 1, 1) = (1, 0, 1) + (2, 1, 0),$$

so we discard $(3, 1, 1)$ to give the set $\{(1, 0, 1), (2, 1, 0)\}$.

The vectors $(1, 0, 1)$ and $(2, 1, 0)$ are linearly independent, so $\{(1, 0, 1), (2, 1, 0)\}$ is a basis for $\mathrm{Im}(t)$.

Thus $\mathrm{Im}(t)$ is a two-dimensional subspace of the codomain $\mathbb{R}^3$; that is, $\mathrm{Im}(t)$ is a plane through the origin with equation

$$ax + by + cz = 0,$$

for some $a, b, c \in \mathbb{R}$.

Since the basis vectors $(1, 0, 1)$ and $(2, 1, 0)$ belong to $\text{Im}(t)$, the values a, b and c satisfy the simultaneous equations

$$\begin{cases} a & + c = 0, \\ 2a + b & = 0, \end{cases}$$

that is, $c = -a$ and $b = -2a$. So $\text{Im}(t)$ is the plane with equation

$$ax - 2ay - az = 0$$

or, equivalently,

$$x - 2y - z = 0.$$

$(\mathbf{b})$ The kernel of t is the set of vectors (x, y, z) in $\mathbb{R}^3$ that satisfy

$$t(x, y, z) = \mathbf{0},$$

that is,

$$(x + 2y + 3z, y + z, x + z) = (0, 0, 0).$$

Thus we are looking for the values of x, y and z that satisfy the simultaneous equations

$$\begin{cases} x + 2y + 3z = 0, \\ y + z = 0, \\ x + z = 0. \end{cases}$$

The last two equations give $x = -z$ and $y = -z$. The first equation is also satisfied for these values of x, y and z.

Assigning the parameter k to the unknown z, we obtain

$$x = -k, \quad y = -k, \quad z = k.$$

So the kernel of t is

$$\text{Ker}(t) = \{(-k, -k, k) : k \in \mathbb{R}\},$$

that is, $\text{Ker}(t)$ is the line through $(0, 0, 0)$ and $(-1, -1, 1)$.

$(\mathbf{c})$ We see that x, y and z satisfy this system of linear equations precisely when

$$t(x, y, z) = (4, 1, 2).$$

Since

$$4 - 2(1) - 2 = 0,$$

it follows from part (a) that $(4, 2, 1)$ belongs to $\text{Im}(t)$. So the system of linear equations has at least one solution.

We know from part (b) that $\text{Ker}(t) \neq \{\mathbf{0}\}$; thus the system of linear equations has infinitely many solutions.

4.15 $(\mathbf{a})$ We follow Strategy 4.1.

We take the basis $E = \{e^x \cos x, e^x \sin x\}$ for the domain V.

We determine the images of these basis vectors:

$$t(e^x \cos x) = e^x \cos x - e^x \sin x,$$
$$t(e^x \sin x) = e^x \sin x + e^x \cos x.$$

The set $\{e^x \cos x - e^x \sin x, e^x \sin x + e^x \cos x\}$ is linearly independent, so it is a basis for $\text{Im}(t)$. The basis has two elements, so $\dim \text{Im}(t) = 2$.

Thus $\text{Im}(t)$ is a two-dimensional subspace of the codomain V. Since V is two-dimensional, it follows that $\text{Im}(t)$ is the whole of V.

$(\mathbf{b})$ Since the dimension of the domain V is 2, it follows from the Dimension Theorem that

$$\dim \text{Im}(t) + \dim \text{Ker}(t) = 2.$$

We know from part (a) that $\dim \text{Im}(t) = 2$. It follows that $\dim \text{Ker}(t) = 0$, that is,

$$\text{Ker}(t) = \{\mathbf{0}\}.$$

Index